Reviews and critical articles covering the entire field of normal anatomy (cytology, histology, cyto- and histochemistry, electron microscopy, macroscopy, experimental morphology and embryology and comparative anatomy) are published in Advances in Anatomy, Embryology and Cell Biology. Papers dealing with anthropology and clinical morphology that aim to encourage cooperation between anatomy and related disciplines will also be accepted. Papers are normally commissioned. Original papers and communications may be submitted and will be considered for publication provided they meet the requirements of a review article and thus fit into the scope of "Advances". English language is preferred.

It is a fundamental condition that submitted manuscripts have not been and will not simultaneously be submitted or published elsewhere. With the acceptance of a manuscript for publication, the publisher acquires full and exclusive copyright for all languages and countries.

Twenty-five copies of each paper are supplied free of charge.

Manuscripts should be addressed to

Co-ordinating Editor

Prof. Dr. H.-W. **KORF**, Zentrum der Morphologie, Universität Frankfurt, Theodor-Stern Kai 7, 60595 Frankfurt/Main, Germany
e-mail: korf@em.uni-frankfurt.de

Editors

Prof. Dr. F. **BECK**, Howard Florey Institute, University of Melbourne, Parkville, 3000 Melbourne, Victoria, Australia
e-mail: fb22@le.ac.uk

Prof. Dr. F. **CLASCÁ**, Department of Anatomy, Histology and Neurobiology,
Universidad Autónoma de Madrid, Ave. Arzobispo Morcillo s/n, 28029 Madrid, Spain
e-mail: francisco.clasca@uam.es

Prof. Dr. M. **FROTSCHER**, Institut für Anatomie und Zellbiologie, Abteilung für Neuroanatomie,
Albert-Ludwigs-Universität Freiburg, Albertstr. 17, 79001 Freiburg, Germany
e-mail: michael.frotscher@anat.uni-freiburg.de

Prof. Dr. D.E. **HAINES**, Ph.D., Department of Anatomy, The University of Mississippi Med. Ctr.,
2500 North State Street, Jackson, MS 39216–4505, USA
e-mail: dhaines@anatomy.umsmed.edu

Prof. Dr. N. **HIROKAWA**, Department of Cell Biology and Anatomy, University of Tokyo,
Hongo 7-3-1, 113-0033 Tokyo, Japan
e-mail: hirokawa@m.u-tokyo.ac.jp

Dr. Z. **KMIEC**, Department of Histology and Immunology, Medical University of Gdansk,
Debinki 1, 80-211 Gdansk, Poland
e-mail: zkmiec@amg.gda.pl

Prof. Dr. E. **MARANI**, Department Biomedical Signal and Systems, University Twente,
P.O. Box 217, 7500 AE Enschede, The Netherlands
e-mail: e.marani@utwente.nl

Prof. Dr. R. **PUTZ**, Anatomische Anstalt der Universität München,
Lehrstuhl Anatomie I, Pettenkoferstr. 11, 80336 München, Germany
e-mail: reinhard.putz@med.uni-muenchen.de

Prof. Dr. J.-P. **TIMMERMANS**, Department of Veterinary Sciences, University of Antwerpen,
Groenenborgerlaan 171, 2020 Antwerpen, Belgium
e-mail: jean-pierre.timmermans@ua.ac.be

203
Advances in Anatomy, Embryology and Cell Biology

Co-ordinating Editor

H.-W. Korf, Frankfurt

Editors

F. Beck, Melbourne · F. Clascá, Madrid
M. Frotscher, Freiburg · D.E. Haines, Jackson
N. Hirokawa, Tokyo · Z. Kmiec, Gdansk
E. Marani, Enschede · R. Putz, München
J.-P. Timmermans, Antwerpen

Sara Gil-Perotín, Arturo Álvarez-Buylla
and José Manuel García-Verdugo

Identification and Characterization of Neural Progenitor Cells in the Adult Mammalian Brain

With 31 Figures

Sara Gil-Perotín
Instituto Cavanilles de Biodiversitat i
  Biologia Evolutiva
Polígono La Coma s/n
46980 Paterna (Valencia)
Spain
e-mail: sara.garcia@uv.es

Arturo Álvarez-Buylla
Neurosurgery Research
10 Kirkham St., Room K-127
San Francisco, CA 94143
USA
e-mail: abuylla@stemcell.ucsf.edu

José Manuel García-Verdugo
Centro de Investigación Príncipe Felipe
Avda. Autopista El Saler, 16,
46013 Valencia
Spain

Instituto Cavanilles de Biodiversitat i
  Biologia Evolutiva
Polígono La Coma s/n
46980 Paterna (Valencia)
Spain
e-mail: j.manuel.garcia@uv.es

ISSN 0301-5556
ISBN 978-3-540-88718-8   e-ISBN 978-3-540-88719-5

Library of Congress Control Number: 2008939433

© 2009 Springer-Verlag Berlin Heidelberg

This work is subject to copyright. All rights are reserved, whether the whole or part of the material is concerned, specifically the rights of translation, reprinting, reuse of illustrations, recitation, broadcasting reproduction on microfilm or in any other way, and storage in data banks. Duplication of this publication or parts thereof is permitted only under the provisions of the German Copyright Law of September 9, 1965, in its current version, and permission for use must always be obtained from Springer-Verlag. Violations are liable to prosecution under the German Copyright Law.

The use of general descriptive names, registered names, trademarks, etc. in this publication does not imply, even in the absence of a specific statement, that such names are exempt from the relevant protective laws and regulations and therefore free for general use.
Product liability: The publishers cannot guarantee the accuracy of any information about dosage and application contained in this book. In every individual case the user must check such information by consulting the relevant literature.

Printed on acid-free paper

9 8 7 6 5 4 3 2 1

springer.com

Acknowledgments

We are grateful to Vicente Hernández Rabaza, Clara Alfaro-Cervelló, and Melissa Lezameta Morgan for providing us with some images, and to Francisco Clascá, Mario Soriano-Navarro, and Maria Duran Moreno for their useful comments.

List of Contents

1	**Historic Overview**	1
2	**Research Methodologies for Adult Neurogenesis.**	5
2.1	Immunohistochemistry	5
2.1.1	Markers of Proliferation	5
2.1.2	Phenotypic Markers	8
2.2	Electron Microscopy	11
2.3	Neurosphere Assay	17
2.3.1	Limitations of the Neurosphere Assay	20
2.4	Fluorescence-Assisted Cell Sorting (FACS) Analysis in Stem Cell Research	20
2.5	Transgenic Animals and the *cre–lox* System	22
2.6	Transplantation of Adult NSCs	23
2.7	Integration and Functionality of Newborn Cells in the Adult Brain	25
3	**Neurogenesis in the Intact Adult Mammalian Central Nervous System**	27
3.1	Description of Neurogenic Regions in the Adult Mammalian Brain	27
3.1.1	Subventricular Zone	27
3.1.2	Hippocampal Dentate Gyrus	42
3.1.3	Concept of Neurogenic Niche	45
3.2	Identification of the Adult Neural Stem Cell in the SVZ	47
3.3	Other Proliferating and Neurogenic Centers in the Adult Brain	49
3.3.1	The Subcallosal Region	50
3.3.2	The Central Canal of the Spinal Cord	50
3.4	Distinct Features of Different Species: Comparative Study of Mice and Humans	53
3.4.1	Bovine Lateral Ventricles	53
3.4.2	Rabbit Lateral Ventricles	55
3.4.3	Primate Lateral Ventricles	56
3.4.4	Human Lateral Ventricles	58
4	**Oncogenesis vs. Neurogenesis**	63
5	**Adult Neurogenesis Under Pathological Stimulation: Ischemia**	67
5.1	Concept and Epidemiology	67
5.2	Effects of Ischemia on the Brain and the SVZ	69

5.3	Extracellular Factors and Neurogenesis After Stroke	73
5.4	Stem Cell-Based Therapies in Ischemia	74
6	**Therapeutic Potential of Neural Stem Cells**	77
7	**Concluding Remarks**	81
	References	85
	Index	103

Abstract

Adult neurogenesis has been questioned for many years. In the early 1900s, a dogma was established that denied new neuron formation in the adult brain. In the last century, however, new discoveries have demonstrated the real existence of proliferation in the adult brain, and in the last decade, these studies led to the identification of neural stem cells in mammals. Adult neural stem cells are undifferentiated cells that are present in the adult brain and are capable of dividing and differentiating into glia and new neurons. Newly formed neurons terminally differentiate into mature neurons in the olfactory bulb and the dentate gyrus of the hippocampus. Since then, a number of new research lines have emerged whose common objective is the phenotypical and molecular characterization of brain stem cells. As a result, new therapies are successfully being applied to animal models for certain neurodegenerative diseases or stroke. This work is being or will be extended to the adult human brain, and so it provides purpose and hope to all previous studies in this field. We are still far from clinical therapies because the mechanisms and functions of these cells are not completely understood, but we appear to be moving in the right direction.

Abbreviations

3HT	Tritiated thymidine
AP	Alkaline phosphatase
APB	AraC plus procarbazol
AraC	Cytosine-beta-D-arabinofuranoside
BDNF	Brain-derived neurotrophic factor
bFGF	Basic fibroblastic growth factor
BMSC	Bone marrow stem cell
BrdU	Bromodeoxyuridine
CC	Corpus callosum
CSF	Cerebrospinal fluid
DAB	Diaminobenzidine
Dcx	Doublecortin
DG	Dentate gyrus
EGF	Epidermal growth factor
EM	Electron microscopy
ENU	*N*-ethyl-*N*-nitrosourea
EPO	Erythropoietin
ESC	Embryonic stem cell
FACS	Fluorescence-assisted cell sorting
GCL	Granular cell layer
GDNF	Glial-derived neurotrophic factor
GFAP	Glial fibrillary acidic protein
GFP	Green fluorescent protein
HB-EGF	Heparin-binding epidermal growth factor
HH3	Phosphorylated histone H3
IGF-1	Insulin-like growth factor-1
LacZ	Beta-galactosidase
LM	Light microscopy
ML	Molecular layer
MRI	Magnetic resonance imaging
NSC	Neural stem cell
OB	Olfactory bulb
OPC	Oligodendrocyte precursor cell

OT	Olfactory tract
PCNA	Proliferating cell nuclear antigen
PDGF	Platelet-derived growth factor
PSA-NCAM	Polysialylated neural cell adhesion molecule
PTEN	Phosphatase and tensin homolog deleted on chromosome 10
RER	Rough endoplasmic reticulum
RMS	Rostral migratory stream
SCR	Subcallosal region
SEM	Scanning electron microscopy
SGZ	Subgranular zone
SVZ	Subventricular zone
Tuj1	Beta-tubulin III
VEGF	Vascular endothelial growth factor
VZ	Ventricular zone

1
Historic Overview

Given the tremendous advance that neuroscience has experienced in recent years, some of the pillars on which it has been sustained have started to collapse. Not so long ago, it was believed that we were born with a fixed number of neurons that died to a greater or lesser extent throughout our lifetime. It was widely accepted that the production of new neurons did not persist after birth in the healthy brain or under pathological conditions. This fact was both recognized and defended by scientists and society as a dogma.

We can now state, however, that adult neurogenesis exists. New neurons are produced during adulthood in all of the vertebrates that have been studied, from fish to mammals, including the human species. In nonmammal vertebrates, it has been described that the degree of neurogenesis is higher and that it affects more regions than in mammals. This fact does not limit the relevance that the presence of neural stem cells in the human brain has to neuroscience. We now know more than we knew decades ago, but we still have to learn how to communicate with stem cells.

The process of learning that neurogenesis actually exists has taken almost a century and has progressed slowly so far. In the late 1800s, scientists worldwide, including the prestigious Spanish researcher, Santiago Ramon y Cajal (1913), maintained that neurogenesis was a process restricted to brain development that ceased after birth. This conclusion was the result of studying the histology of the brain with the techniques of the time, such as Nissl and silver impregnation. Most researchers defined neurons as cells that were characterized by the presence of dendritic arborizations. When dendrites were not well developed, cells were thought to be in the process of differentiation, plastic changes, or the result of a histological artifact.

In the first half of the twentieth century, however, occasional studies described these less differentiated neurons as potential newly formed neurons. This hypothesis was speculation that lacked rigorous scientific work to support it. A remarkable study by Sugita (1918) described an increase in new neurons in the rat cerebral cortex after comparing neonates with 20-day postnatal animals. This result has since been attributed to stereological errors in his analysis that disregarded the increase in brain volume with age.

The observation of mitosis correlates with the existence of active proliferating centers. In the past, however, some authors identified mitosis in the brain as dividing glial cells or endothelium. Occasionally mitotic figures were observed in the adult lateral ventricles, regions that play an important role during embryonic development. After a few decades, an elegant study drew the interest of scientists back to mitosis (Bryans 1959). This study was performed in rats in which colchicine was injected. Colchicine is a substance that binds to microtubules and disrupts the mitotic spindle, arresting cells in mitosis. After injection, a large amount of mitosis was found in the cells surrounding the lateral ventricles, making it possible for these ventricular walls to be a source of new neurons in the adult brain. Although scientists were on the right track, technical resources were limited at that time, and it was not possible to prove that mitotic cells gave rise to new neurons. This fact led the scientific community to overlook these new findings.

However, in the early 1950s, specific markers of cell proliferation—such as tritiated thymidine—were discovered, and research into adult neurogenesis revived. Tritiated thymidine (3HT) incorporates into nucleic acid during its synthesis in the S-phase of the cell cycle that precedes cell division, and it can be detected later by autoradiography. Altman (1962, 1969a, b), a self-taught postdoctoral student, was a pioneer in the use of this substance in the adult brain, and he published a series of reports showing neurogenesis in diverse brain regions of the young and adult rat brain. Animals were sacrificed shortly after injection in order to label proliferating centers, and at longer times after injection to study the fate of the cells that proliferated in these centers. Altman described neurogenesis in the neocortex, hippocampus, and olfactory bulb, and went on to not only identify the newly formed cells as microneurons, but also to relate neurogenesis to a potential role in memory and learning. Many scientists felt that this type of neurogenesis should be interpreted as being sporadic and residual, i.e., a peculiarity of the rodent brain. Although his work was accurate to a great extent, he received criticism for two reasons: (1) there was no proof of the existence of migration routes from the germinal zones (lateral ventricle wall) toward the final locations of other brain areas, and; (2) the micronuclei—supposedly neuronal—may have been confused with the nuclei of glial cells. Altman had no proof of cell migration pathways other than that toward the olfactory bulb, and he did not know how to functionally explain the meaning of the generation of new neurons. We should not forget the work done in this field by Bayer, who studied neurogenesis by means of the incorporation of 3HT into the third ventricle, the septal nucleus, the hippocampal region, amygdalae, the superior colliculus, and the olfactory bulb (Altman and Bayer 1979a, b, c, 1981a, b; Bayer 1980a, b, 1983). The majority of his work, however, was performed during brain development. Another supporter of adult neurogenesis was Kirsche (1967). By using 3HT, Kirsche demonstrated the existence of active proliferation sites (sulci) in the adult brains of several groups of vertebrates. He attempted to perform a comparative study of homologous regions, but his work was not well accepted. This was because he believed that the brain was immature in nonmammalian vertebrates and had not yet undergone complete growth.

Further interest in neurogenesis can then be attributed to a technical advance, the electron microscope. Electron microscopy (EM) allowed the observation of cells at high resolution, enabling neurons and glial cells to be differentiated. Using EM, Kaplan studied cells labeled with 3HT in the rodent brain, and confirmed the existence of adult neurogenesis and the presence of synaptic contacts in the newly formed neurons (Kaplan 1981, 1985; Kaplan et al. 1985). He also demonstrated long survival times for these young neurons. These studies were performed in the hippocampus, an area that was becoming increasingly important because of its involvement in learning and memory.

The scientific community started to wonder whether this phenomenon could be extended to all vertebrates and mammals, especially to humans. A number of reports showed that, although neurogenesis is not restricted to mammals, it is present in all vertebrates, and the number of proliferative centers involved is higher the closer they are to the base of the phylogenetic tree. At present, we know that neurogenesis occurs in fish in the whole brain, particularly in the cerebellum, while it is exclusive to the telencephalon in all other vertebrates. Only in amphibians and certain ophidian reptiles can new neurons be found in the optic tectum, probably owing to their relationship with the retina.

By this time, the existence of adult neurogenesis in nonhuman mammals, at least in the dentate gyrus of the hippocampus and the olfactory bulb, was widely accepted. The main concern was elucidating whether adult neurogenesis existed in the human brain and where it occurred. To answer that question, and given the lack of human brain material, studies were performed in species that were closely related to humans, such as primates. In a series of detailed studies using 3HT, Rakic (1985a, b) concluded that there was no adult neurogenesis in primates and possibly not in humans either. He did not find labeled cells in the adult primate brain. He hypothesized that neurogenesis in the brains of lower vertebrates could be the result of immaturity of the brain after birth and the need for new neurons to complete the number of neurons necessary to reach the adult state. Rakic also used physiology to explain this event. New neurons of lower vertebrates could store new data throughout life, thus enabling the learning process. In contrast, and as a result of evolution, primates would be born with mature neuronal units that are ready to store all information after birth. The arrival of newly formed neurons may alter not only these systems but also the existing circuitry, causing the loss of stored data. This hypothesis was accepted up until a few years ago.

During this same period, a satisfactory explanation was given to address the physiological meaning of adult neurogenesis. Fernando Nottebohm (Goldman and Nottebohm 1983; Nottebohm 1985; Paton and Nottebohm 1984), while working with songbirds, demonstrated that adult neurogenesis occurred in a telencephalic center that was directly related to vocal learning. Vocal learning is essential for communication between birds, reproduction and the establishment of territories. He showed that: (1) there were several centers involved, but that only the high vocal center showed neurogenesis; (2) the newly formed neurons received synaptic contacts, their ultrastructure was similar to that in mature neurons, and they were functional; (3) when eliminated, birds were impaired in the learning of new sounds, and; (4) the regions involved were four times larger in males than in females, and this phenomenon was reverted by injecting testosterone into females. This was accompanied by an increase in neurogenesis in females that was not observed in untreated females.

A few years later, migratory cells were described in the adult reptile brain (Garcia-Verdugo et al. 1986). These cells were located between the lateral ventricles and a region of neuronal cell bodies called the medial cortex. The medial cortex of reptiles has been compared to the dentate gyrus in mammals. By using 3HT and EM, the existence of neurogenesis was also demonstrated in reptiles (Lopez-Garcia et al. 1988). Newly formed neurons in the reptilian brain were born close to the ventricles and migrated radially toward the majority of telencephalic regions. After selective chemical injury, and by increasing neurogenesis, the reptilian cerebral cortex was repaired completely (Font et al. 1997).

Despite the fact that the field of neurogenesis in birds and reptiles advanced rapidly, limited studies were performed on neurogenesis in mammals. However, the use of new cell markers and novel techniques questioned Rakic's hypothesis, as several studies have demonstrated postnatal neurogenesis in primates (Gould et al. 1998, 1999a). A surprising study by Gould et al. (1999a) described neurogenesis in associative regions in primates for the first time. In a recent review of the topic, and after considerable controversy, this work was thoroughly reviewed (Gould 2007). Simultaneously, adult neurogenesis was also found in the human brain. The first demonstration in humans was performed in five cancer patients who were injected with a proliferation marker, bromodeoxyuridine (BrdU), which works similarly to 3HT, for the purpose of determining tumor activity (Eriksson et al. 1998). The brains of these patients were studied after death, and labeled cells were detected in the hippocampal dentate gyrus. Specific neuronal immunohistochemistry markers also stained these cells.

All of this data has aroused the curiosity of the scientific community, and increasing effort has been expended to finally demonstrate the validity of adult neurogenesis. Elucidating the identity of the cells responsible for neurogenesis will permit us to understand the physiological reason for the generation of new neurons, to establish dialogue with them, and therefore to direct them to damaged regions, undoubtedly leading to a fascinating future for this field of research.

2
Research Methodologies for Adult Neurogenesis

As in any field of science, progress in neuroscience has been, in part, the result of a multidisciplinary and technological development that has enabled us to recognize biological events from observations which were, until then, uncertain.

The first works in neurogenesis were based on anatomical and morphological descriptions of the mitotic figures that were either found randomly within the brain or concentrated in definite regions. The emergence of electron microscopy (EM), which enabled greater magnifications, implied a significant gain of knowledge compared to the information obtained with light microscopy (LM). However, the exclusive use of EM or LM to study proliferation did not prove the existence of neurogenesis. It was necessary to combine morphology with immunohistochemistry, cell culture, electrophysiology, and molecular genetics to finally demonstrate that new cells were born in the adult brain and that these cells were able to become functional afterwards. In vitro technologies enable the growth, amplification and isolation of cells in terms of their immunochemical pattern and morphology by using flow cytometry. In addition, the use of transgenic animals to amplify or knock out target genes conditionally or permanently is highly significant for studies of the functions of gene products. Here, we briefly list and describe some of the methods currently used to analyze adult neurogenesis.

2.1
Immunohistochemistry

2.1.1
Markers of Proliferation

There are two main classes of markers to label proliferation: exogenous and endogenous markers. Endogenous markers are molecules that the cell expresses during the progression of the cell cycle, and they correlate with the duplication of its DNA or with the mitotic division. This type of marker has been widely used in human tissue given the ethical and technical limitations on the injection of exogenous markers that bind to DNA in vivo and may produce DNA mutations. Among this group, Ki-67 (Fig. 1a) and PCNA (not shown) are worth mentioning.

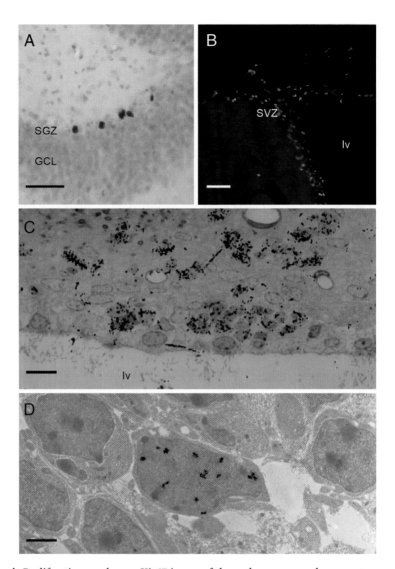

Fig. 1a–d Proliferation markers. **a** Ki-67 is one of the endogenous markers most commonly used to measure proliferation, especially in human samples. In this picture of the hippocampal fascia, Ki-67+/DAB-labeled cells can be observed. The fascia is counterstained with Nissl stain. Labeled cells are located in the subgranular zone (SGZ), one of the proliferation centers in the adult brain, *scale bar* 50 µm. **b** An exogenous marker, bromodeoxyuridine (BrdU), is injected into the animal with variable doses and survivals. This permits differentiation or proliferation studies. In this case, the injected dose was 100 µg g^{-1} of body weight and the survival time was 1 h (labeling cells in the S-phase). The figure shows a section of the subventricular zone (SVZ) and the labeling technique is immunofluorescence, *scale bar* 120 µm. Another widely used exogenous marker is tritiated thymidine (3HT) (**c,d**). **c** The semithin section depicts many labels in the SVZ after autoradiography (*black dots*). Sections were counterstained with toluidine blue. In this experiment, the animal was infused with 3HT-thymidine via an osmotic pump for 24 h, and it was sacrificed three days later, *scale bar* 10 µm. **d** Detail from electron microscopy (EM) in which a cell is labeled with 3HT-thymidine in the neonatal SVZ, *scale bar* 2 µm (*GCL*, granular cell layer; *lv*, lateral ventricle)

Ki-67 is a nuclear protein present in all the phases of the cell cycle except in G0, thus providing a measure of cells that have entered the cell cycle (Gerdes et al. 1983, 1984). Although less frequently used, PCNA (proliferating cell nuclear antigen), an auxiliary protein for the DNA polymerase delta, also allows the assessment of cell proliferation (Bravo et al. 1987; Takasaki et al. 1981). Its expression peak coincides with the G1/S restriction point, and decreases through G2 (Kurki et al. 1986; Takasaki et al. 1981). Other markers, although much less used, are P34cdc2, first identified in HeLa cells that regulate the G2/M transition and initiation of mitosis (Draetta and Beach 1989; Draetta et al. 1988), and phosphorylated histone H3 (HH3), which is expressed in the G2 and M phases of the cell cycle (Bosch et al. 1993; Chou et al. 1990; Wong et al. 1990). The advantage of endogenous markers is that we do not introduce molecules into the cell that might alter the normal cell functioning. Their main disadvantage is the difficulty involved in following the temporal fate of cells after proliferation has occurred.

In contrast, exogenous markers have to be added to cells and are incorporated into the nucleic acid in the S-phase. These markers allow us to follow up the cell lineage from a neural stem cell (NSC), or precursor, to the mature neuron, including differentiation and survival studies. The two thymidine analogs, bromodeoxyuridine (BrdU) (Fig. 1b) and tritiated thymidine (3HT) (Fig. 1c,d), constitute the most important exogenous markers (Dolbeare 1995; Dolbeare and Selden 1994; Kriss and Revesz 1961), and were the key players in the discovery of adult neurogenesis, as mentioned before (Altman 1969a). BrdU was introduced to study cell proliferation in the developing nervous system according to Nowakowski et al. (1989).

The administration protocol depends on the experiment, but injections are normally administered intraperitoneally or dissolved in drinking water. The dose varies between 50 and 200 mg kg^{-1} day^{-1} in the case of BrdU, and from 5 to 15 mCi kg^{-1} for 3HT. Some authors consider that lower doses are not sufficient to label all of the proliferating cells, while others consider that higher doses might kill the cells (Gould and Gross 2002). Labeling the total proliferating population of cells in the brain would require multiple injections every 2 h throughout the cell cycle (Nowakowski et al. 1989).

These exogenous markers have their advantages and disadvantages. 3HT is radioactive and takes at least a month before samples can be developed by autoradiography and, therefore, to obtain results. The autoradiography process maintains the ultrastructure of the cell, allowing us to visualize it with EM. 3HT has been abandoned in favor of BrdU because of it is easier to use and handle, although its carcinogenicity has to be taken into account. The process of visualizing BrdU requires a DNA denaturation with hydrochloric acid that deteriorates the tissue structure; however, it is a quick method and permits colocalizations of BrdU to be studied with specific cell markers (Table 2.1). The main concern over the use of 3HT or BrdU is that the results depend on a number of factors such as the injection protocol, the length of the S-phase of the cells (12–24 h for NSCs), marker exposure, and cell survival (Hayes and Nowakowski 2002).

Given the pitfalls and limitations of previous methods, other approaches have been investigated that might provide additional information about proliferation. One such case is the use of retroviruses. A requirement for retroviral integration

Table 2.1 Basic immunochemistry protocol

- *Fixatives*: 4% paraformaldehyde (PFA)
- *Reagents*:
    1. Blocking buffer: 10% normal serum in 0.1 M phosphate buffer (PB)
    2. 3% hydrogen peroxide solution in distilled water
    3. ABC kit: to form biotin–avidin complexes with the secondary antibodies (commercialized kit)
    4. DAB solution: 0.5 mg mL^{-1} diaminobenzidine in 0.1 M PB. Add 5 µL of freshly prepared 0.3% hydrogen peroxide to the solution just before use
- *Protocol*:
    1. Block nonspecific binding with 10% NGS for 30 min
    2. Incubation with primary antibody overnight at 4°C
    3. Wash in 0.1 M PB (3 × 10 min)
    4. Incubation with secondary antibody for 1 h at room temperature (RT)
    5. Wash in 0.1 M PB (3 × 10 min)
    6. ABC incubation for 30–45 min
    7. Develop with DAB solution
- BrdU staining: pretreatment with 0.1 N ClH for 20 min at 37°C and 0.1 M sodium borate (pH 8.0) for 10 min at RT. Then follow protocol from step 1

and the expression of viral genes (and target genes) is that target cells should be dividing. This method is very specific because viruses are injected locally (Cheng et al. 1996). This means that retroviruses are perfect for introducing markers such as green fluorescent protein (GFP) into proliferating cells, thus allowing for direct in vivo observation. It is also possible to measure proliferation in vitro using flow cytometry of these GFP$^+$ cells (see Chap. 2.4). A disadvantage of retroviral techniques is the relative frequency of gene silencing some time after infection.

2.1.2
Phenotypic Markers

The demonstration of adult neurogenesis was not only sustained by the presence of dividing cells, as these might have been reactive astrocytes or microglia. It was essential to prove that those cells undergoing mitosis were able to become neurons. This has been possible to achieve mainly through the use of antibodies against specific cell markers. The following markers are commonly used to phenotype adult NSCs and neurons.

(a) *Precursor cells*. So far, no exclusive marker to stem-like cells has been identified, which reflects the controversy regarding the identity of stem cells. Nestin is one of the most widely used; it defines a class of intermediate filament proteins, and was found to be specifically expressed in neural progenitor cells (Lendahl et al. 1990). Unfortunately, it is not a specific marker for the distinct subpopulations of progenitors. In the adult subventricular zone (SVZ) and dentate gyrus (DG), the cells identified as NSCs also expressed, but obviously not specifically, the astrocytic marker for the glial fibrillary acidic protein (GFAP) (Fig. 2a) (Doetsch et al. 1999). In addition, Sox-2, a HMG box transcription factor, has also been identified as a marker for multipotential NSCs (not shown) (Brazel et al. 2005; Ellis et al. 2004). It is expressed in the adult SVZ (Ferri et al. 2004), and plays an important role in NSC maintenance (Episkopou 2005; Ferri et al. 2004). However, it is still to be elucidated whether it is a putative marker for precursors or for bona fide stem cells. Our experience indicates that it is a marker that is widely expressed by all types of precursors in the SVZ and is therefore not useful as a unique stem cell marker. In the search for a more specific marker, an intracellular marker, mouse Musashi1 (m-Msi1) (Fig. 2b)—which is a neural RNA-binding protein—was identified, and its expression was restricted to proliferating neuronal and/or glial precursor cells during development (Kaneko et al. 2000; Sakakibara et al. 1996), and in the postnatal brain (Sakakibara and Okano 1997). Interestingly, newly generated postmitotic neurons and oligodendrocytes did not express it. m-Msi1 is a candidate for a marker of "stemcellness," but it cannot be considered to be isolated, as there are m-Msi1⁻ immature precursors that meet all of the requirements needed to be stem cells. CD133, a novel marker that has been related to brain cancer stem cells, among others, is not present in neurogenic astrocytes in the SVZ, but in embryonic neural stem cells, ependymal cells, and glioblastoma cells (Pfenninger et al. 2007).

(b) *Immature neurons.* A widely used marker for migrating neuroblasts or immature neurons is Tuj1, the neuron-specific class III tubulin, a cytoskeletal protein that is expressed in all postmitotic neurons (Fig. 2c) (Menezes and Luskin 1994; Thomas et al. 1996). Doublecortin (Dcx), which encodes a microtubule-associated protein, colocalizes with Tuj1 in the migrating neuroblasts in the adult telencephalon (Fig. 2d) (Nacher et al. 2001; Yang et al. 2004), as does PSA-NCAM (polysialylated neuronal cell adhesion molecule) (not shown) (Rousselot et al. 1995). The mammalian RNA-binding protein, Hu—the homolog of the *Drosophila* neuron-specific RNA-binding protein, Elav—is expressed exclusively in postmitotic neurons and is considered an early marker (Barami et al. 1995). Dlx2 is a transcription factor involved in regulating Notch signaling, among other factors, which in turn mediates the temporal control of subcortical telencephalic neurogenesis in mice during development and in adulthood. It stains immature neurons and more uncommitted precursors. This ambiguity makes data interpretation difficult, and the use of several markers to characterize the progenitors and precursor cells is necessary most of the time.

(c) *Mature neurons.* The most frequently used specific markers for mature neurons are neuron-specific enolase (NSE), neuronal-specific nuclear protein (NeuN) (Fig. 2e), and microtubule-associated protein (MAP-2). In differentiation studies, it is useful to follow up the BrdU⁺ cells that have migrated after dividing, and to terminally acquire the neuronal phenotype in its final destination.

Figure 3 summarizes the more widely accepted markers for each cell subtype.

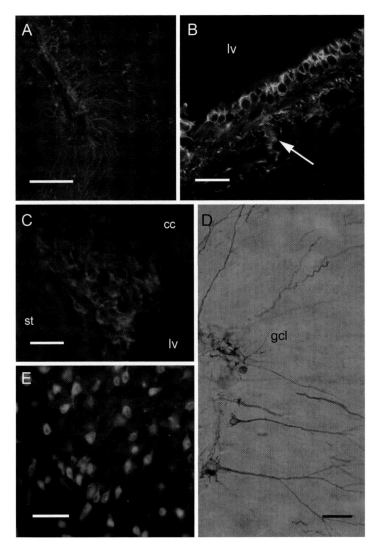

Fig. 2a–e Phenotypic markers. **a** In the horns that the SVZ forms ventrally and dorsally, the density of cells increases, as they are exit points for migration toward the olfactory bulb (OB). In the figure, a ventral horn from the SVZ of adult mice stained with GFAP (*red*) not only shows an increased density in this area, but also shows how astrocytes send expansions toward the brain parenchyma. GFAP stains the cell body and expansions; *scale bar* 50 μm. **b** Musashi (Msi1), an RNA-binding protein, is expressed only by precursor cells during development and in adulthood. The picture shows a Msi1[+] cell (*arrow*) in a section of the SVZ of *Macaca fascicularis* which is also stained with a polyclonal glial fibrillary acidic protein (GFAP) antibody (*green*) that labeled astrocytes and ependymal cells, *scale bar* 25 mm. **c** A widely recognized marker for immature neurons is Tuj1 (*red*) that stains the cell body and expansions of neuroblasts, concentrated in the dorsal horn of the SVZ, *scale bar* 25 mm. **d** Doublecortin (Dcx) is an alternative marker for neuronal precursors and it colocalizes with Tuj1. It is very useful for studying hippocampal neurogenesis, and can be used to measure the maturity of newly formed neurons, *scale bar* 25 μm. **e** NeuN is a marker for mature neurons. This picture shows immunofluorescent staining of NeuN[+] neurons in the brain parenchyma, *scale bar* 25 μm (*st*, striatum; *lv*, lateral ventricle)

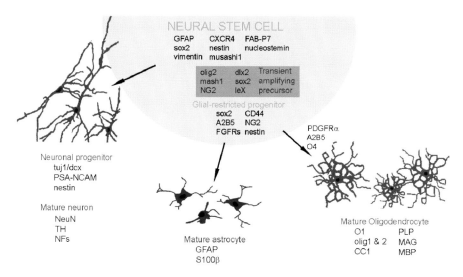

Fig. 3 Phenotypic markers for adult neural stem cells. The inability to find a specific marker for the bona fide neural stem cell and/or its progeny makes it necessary to use a wide range of markers that define one cell type or another in further studies. In the diagram, some of these markers are depicted for the three neural lineages arising from the SVZ

2.2
Electron Microscopy

Because of the unreliability of specific molecular markers in the adult NSCs and their progeny, the main advantage of EM has been characterization using morphological criteria. EM has proven to be a useful tool for characterizing distinct lineages in the proliferative regions of the adult brain (Doetsch et al. 1997; Garcia-Verdugo et al. 1998). Ultrastructural features have allowed scientists to classify the cells contained in the two main neurogenic niches: the SVZ that surrounds the lateral ventricles and the DG of the hippocampus (Doetsch et al. 1997). Furthermore, there are aspects of cells that LM or immunohistochemistry are unable to detect, but which are observable at higher resolutions. One such example is the fine astrocytic expansions that surround migrating neuroblasts in the rostral migratory stream (RMS) or DG (Seri et al. 2001). Another great contribution to the study of neurogenesis is the revelation of empty spaces between the chains of migratory neuroblasts. These spaces were undetectable with LM, and have been very useful for interpreting migration, as they might be the result of (or the condition for) cell movement (Jankovski and Sotelo 1996). In addition, certain observations in cells observed by EM have yet to be described or addressed to date, such as electron-dense particles in the basal corpuscle of the primary cilia or endocytic phenomena between stem cells and daughter cells. More recently, it has been noted that B cells (see the full explanation in Chap. 3) usually contact the ventricular lumen through thin expansions. These expansions cannot be observed with LM, and are confused with adjacent ependymal cells. The protocol to prepare tissue samples for EM is described in Table 2.2.

Table 2.2 Transmission electron microscopy (TEM)

- *Fixatives*: 2% PFA–2.5% glutaraldehyde (GA)
- *Reagents*:
    1. Distilled water
    2. Ethanol (70°, 96°, 100°, dry 100°) (EtOH)
    3. Propylene oxide
    4. Resin (Araldite)
    5. 2% osmium solution (in distilled water)
    6. 7% uranyl acetate (in 70% ethanol)
    7. Lead citrate: Reynolds' solution: 1.33 g lead citrate + 1.76 g sodium citrate + 2 g 1 N NaOH in 50 mL distilled water
- *Protocol*:
    1. Incubate tissue in 2% osmium for 1 h at RT and in the dark
    2. Wash with dH_2O
    3. Progressively dehydrate the tissue with crescent concentrations of EtOH
    4. Incubate for 2 h in 7% uranyl acetate
    5. Dehydrate and treat with propylene oxide for 10 min
    6. Transfer tissue to araldite and shake overnight at RT
    7. Transfer to polypropylene molds and allow to dry at 37°C for 3–4 days
    8. Trim EM blocks and cut 0.5 µm thick sections
    9. Cut ultrathin sections at 60–90 nm
    10. Collect sections on formvar–carbon-coated nickel grids
    11. Stain with Reynolds' solution

Fig. 4a–f EM techniques. **a** This picture shows the result of DAB pre-embedding immunofluorescence against the GFAP labeling of the cytosol of an astrocyte (*asterisk*). After immunochemistry of 50–100 µm sections for GFAP (primary antibody), sections were embedded and cut into semithin sections, then processed for EM, *scale bar* 2 µm. The label is an electron-dense precipitate (*arrow*). **b** Immunochemistry pre-embedding with gold-conjugated antibodies against green fluorescent protein (GFP). After immunochemistry, embedding and sectioning were performed. Then the tissue containing labeled cells was processed for EM observation. Electron-dense dots (*enhanced silver grains*) are a characteristic of the target tissue, *scale bar* 2 µm. In the subset, a dendrite is labeled with gold particles and it establishes synaptic contacts with other dendrite (*arrows*), *scale bar* 200 nm. **c** Postembedding immunochemistry against GFAP where gold particles can be seen to be attached to the glial intermediate filaments (*arrows*), *scale bar* 1 µm. **e** and **f** show LacZ labeling. **d** The SVZ astrocyte that has incorporated iron particles injected into the lumen of the ventricle. When the labeled astrocyte behaves like a neural stem cell, its progeny in the OB are granular neurons, which still contain iron inside the cytosol. Iron can be seen in the semithin section (not shown) as a golden particle and as electron-dense material within the cell under EM (*arrows*), *scale bar* 2 µm. **e** Semithin section showing cells labeled for LacZ (*blue in the square box*). These cells are neurons that form after grafting stem cells from a transgenic animal (expressing LacZ under the specific neuronal enolase promoter). These stem cells differentiate into neurons in the target tissue, and then express LacZ, which is processed to stain cells blue, *scale bar* 15 µm. **f** Detail from the previous picture (*black box*). EM shows electron-dense LacZ deposits within the neuron. LacZ normally localizes in the RER (*arrow*), and perinuclearly within the cell, *scale bar* 5 mm

Electron Microscopy

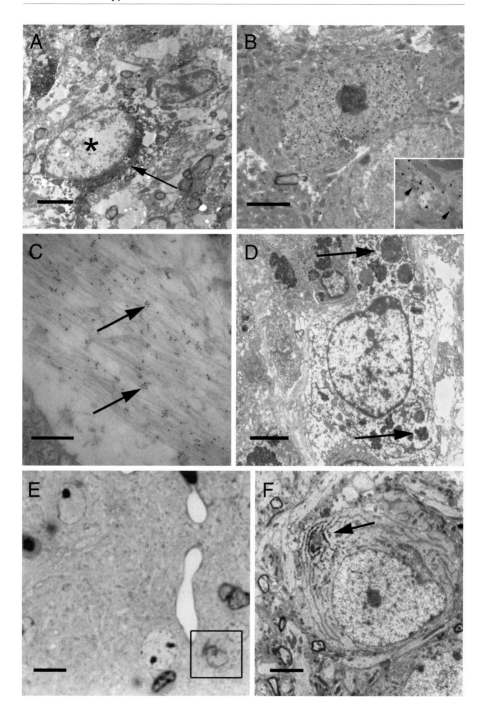

Initially, EM was synonymous with morphology, but after the appearance of immunohistochemistry and proliferation markers, the range of use and the possibilities associated with EM multiplied. This was the case for pre-embedding immunohistochemistry, which allows the localization and identification of prelabeled cells with high resolution, as well as the sublocalization of the marker within the cell. This is possible because the DAB precipitates and its precipitated salt is electron-dense and persists after a "softer" embedding protocol (Fig. 4a). One variety of pre-embedding immunochemistry is the "immunogold" technique (Table 2.3), which uses gold-conjugated secondary antibodies, and electron-dense labeling is cleaner and well defined (Fig. 4b,c).

Since stem cells are astroglial in nature, they have phagocytic properties. Labeling with iron (after phagocytosis) facilitates the study of the migration routes of stem cell progeny (Fig. 4d). It offers another, more interesting, advantage too; the possibility of following labeled cells by magnetic resonance imaging (MRI) in a living animal (Shapiro et al. 2006). Markers that associate with a target gene are very useful in EM. One of the most commonly used markers is the product of the *LacZ* gene (which encodes for beta-galactosidase) under the promoter of the target gene. Beta-galactosidase, when added to a substrate (X-gal), forms deposits around the cell nucleus that are blue-green with LM and electron-dense with EM after embedding.

Table 2.3 Gold immunostaining for EM

- *Fixatives*: 4% PFA
- *Reagents*:
    1. 0.1 M PBS
    2. 0.1 M glycine
    3. 0.2 M sucrose solution
    4. Resine (Araldite)
    5. Propylene oxide
    6. TBS-Tween 20 solution
- *Protocol*:
    1. Fix with 4% PFA in 0.1 M PBS, pH 7.4 overnight
    2. Incubate in 0.1 M glycine solution for 20 min
    3. Incubate in 0.2 M sucrose solution
    4. Process for EM until ultrathin sections have been obtained
    5. Block nonspecific binding with 10% NGS for 30 min
    6. Incubation with primary antibody for 2 h at RT
    7. Wash in TBS-Tween 20
    8. Incubation with secondary biotinylated antibody for 1 h at RT
    9. Wash in TBS-Tween 20
    10. Incubate with gold-conjugated streptavidin for 1 h
    11. Wash with dH_2O
    12. Stain with uranyl acetate and Reynolds' solution

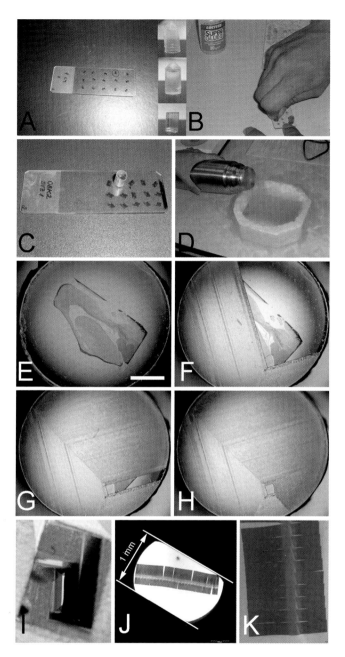

Fig. 5a–k Semithin section after detachment from the glass slide and attachment to the araldite block. **a–d** The semithin section is glued onto the araldite block to obtain ultrathin sections. Liquid nitrogen is required to detach the section from the slide. **e–h** To crop the semithin section into a smaller section, a glass knife is used at a specific angle. The tissue around the section is trimmed while the surface of interest protrudes, *scale bar* 5 mm. **i** Picture showing the particular trimming needed to obtain serial ultrathin sections. **j,k** Micrograph of serial ultrathin sections used to perform three-dimensional reconstructions

This permits a direct correlation to be established between LM and EM, and therefore identifies the cells that express a certain gene (Fig. 4e,f). In the future, the combination of all of these methodologies will help to establish the lineage progression sequence from stem cells to differentiated cells.

Any tissue stained with pre-embedding procedures and cut into semithin sections can be transferred and re-embedded (Fig. 5a–e) for further processing. This allows the study of 3HT-labeled cells in semithin sections (see the protocol in Table 2.4). After autoradiography for 3HT, semithin sections can be glued to araldite blocks (Fig. 5c), and detached from the glass slide by repeated freezing (in liquid nitrogen) and thawing (Fig. 5d). The block with the flat semithin section was mounted onto the ultramicrotome. Ultrathin sections were cut with a diamond knife and examined under EM to determine which cell types incorporated 3HT.

EM-derived ultrathin sections are extremely thin (ranging from 50 to 70 nm). This is a drawback, as cellular structures are larger than 50 nm, and so three-dimensional morphology cannot be observed in a single ultrathin section. To overcome this problem, several serial sections can be cut and set in the grid (with a surface of 2×1 mm), and as many as 100 sections can fit in just one grid. To do this, we have to focus on the regions of interest. This is the most complex stage of the process, because a large surface would need more grids to study, whereas small surfaces will be at risk of leaving the cell being studied out of the analysis. Normally, a semithin section is cropped (Fig. 5e–h), leaving a small surface. In such cases, the borders have to be cut sharply so that sections do not stack together. Several grids will allow us to study the three-dimensionality of a cell or organelle (Fig. 5i,j).

Scanning electron microscopy (SEM) is a useful tool when attempting to study cell surfaces. The single cilium that B cells (astrocytes from the SVZ) sent toward

Table 2.4 Autoradiography for 3HT-thymidine

- *Dose*: Mice were administered a 50 mL injection of 1 mCi [3H] thymidine intraperitoneally
- *Survival times*: depend on the experiment
- *Reagents*:
    1. Autoradiographic emulsion
    2. Photographic developer and fixative
    3. See reagents for TEM processing
- *Protocol*:
    1. Serial 1.5-mm-thick semithin sections
    2. Dip in autoradiographic emulsion and expose for 4 weeks in the dark at 4°C
    3. Develop and fix, then counterstain with 1% toluidine blue
- A cell is considered labeled if six or more silver grains overlaid the nucleus and the same cell was labeled in three adjacent sections
- 3HT-labeled cells identified in the semithin sections can be selected for electron microscopic examination

the lumen of the ventricle could be observed using SEM (**Fig. 6; Table 2.5**). SEM can also be employed to study ultrathin sections (Micheva and Smith 2007). This technique, combined with immunofluorescence, enables us to relate molecular markers to ultrastructural analysis.

2.3
Neurosphere Assay

Stem cells have been classically defined by their behavior in vitro (Potten and Loeffler 1990). NSCs are the least committed cells of the nervous system, and differ from the neural progenitor or precursor cells, which have a more restricted differentiation potential. Neurogenic regions contain both types of cells. Due to scientific reasons and the lack of ideal markers, it is necessary to apply operational criteria to separate these populations. The functional properties of NSCs as stem-like cells are: (1) multipotency, i.e., the ability to yield mature cells in all three fundamental neural lineages throughout the nervous system: neurons, astrocytes, and oligodendrocytes; (2) repairing potential, i.e., the ability to populate a developing

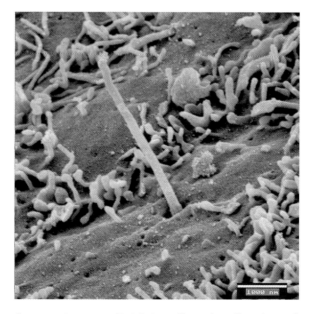

Fig. 6 Scanning electron microscopy (SEM). SEM allows the cell surface to be studied. In the SVZ, the ependymal layer is composed of ependymal cells with abundant cilia and microvellosities in the apical pole. In the picture, a single cilium arises from the cell surface, *scale bar* 1 μm. This is characteristic of another cell type, B cells, as we will explain later in Chap. 3

Table 2.5 Scanning electron microscopy (SEM)

- *Reagents*:
    1. 2% osmium in 0.1 M PB
    2. 0.1 M PB solution
    3. Distilled water
    4. EtOH (25, 50, 70, 95, and 100%)
- *Protocol*:
    1. Rinse tissue in PB (2 × 10 min)
    2. Transfer tissue to 2% osmium for 1–2 h. Keep in the dark at RT
    3. Rinse in distilled H_2O at 4°C (5 × 10 min)
    4. Rinse tissue in 25% EtOH for 10 min at 4°C
    5. Rinse tissue in 50% EtOH for 10 min at 4°C
    6. Rinse tissue in 70% EtOH for 10 min at 4°C (it can be stored in 70% EtOH for 1 week at 4°C)
    7. Rinse tissue in 70% EtOH for 10 min at 4°C
    8. Rinse tissue in 95% EtOH (3 × 10 min at 4°C)
    9. Rinse tissue in 100% EtOH (3 × 10 min at 4°C)
    10. Critical point drying
    11. Sputter coating
    12. Scanning

region and/or repopulate an ablated or degenerated region of the nervous system with the appropriate cell types; (3) the ability of to be serially transplanted, and; (4) self-renewal, i.e., the ability to produce daughter cells or new NSCs with identical properties and potential.

For the first time, Reynolds and Weiss (1992) reported the existence of a type of cells in the adult mammalian brain that had self-renewal properties. They cultured striatum-derived cells from the adult rodent brain in the presence of the mitogenic factor EGF (epidermal growth factor). In the culture plate, they proliferated in clusters to form sphere-like structures. When dissociated and replated, these cells gave rise to new clonally derived neurospheres (secondary neurospheres). This was the basis for the neurosphere assay. Initially it was thought that neurospheres were derived directly from the striatum (Reynolds et al. 1992), but subsequent studies demonstrated that the source of the cells, which behave as stem cells in vitro, is the SVZ that underlies the striatum (Lois and Alvarez-Buylla 1993). Culture conditions were thoroughly analyzed and another mitogenic factor, bFGF (basic fibroblast growth factor), proved to be sufficient for neurospheres to grow (Vescovi et al. 1993). An interesting fact is that the presence of both factors in the human brain is required for stem cell propagation, while each factor separately allows the neurosphere to grow in the rodent (Vescovi et al. 1999). The protocol of the SVZ-derived cells in culture is shown in Table 2.6.

Table 2.6 Neurosphere culture

- *Reagents*:
  1. DMEM–F12 1×
  2. Hormone mixture (B7 or N2 supplement)
  3. 2 mM L-glutamine
  4. Antimitotic/antibiotic
  5. Heparin
  6. Growth factors: EGF, bFGF
- *Protocol*:
  1. Extract the SVZ from fresh brain and put on DMEM–F12 1× on ice
  2. Cut into small pieces and triturate using a fire-polished Pasteur pipette
  3. Plate in DMEM–F12 with 20 ng mL^{-1} EGF and 10 ng mL^{-1} FGF
  4. For differentiation studies, plate in polyornithine-coated chamber slides without mitogens in the presence of 1% FBS

The property of self-renewal is dependent on cell development. That is, the symmetric division of neuroepithelial cells and radial glia predominates during embryogenesis. At midgestation, the SVZ arises and the asymmetric division increases to gradually allow the formation of the mature structures that will constitute the adult brain. Thus, the potential of NSCs for self-renewal changes as development proceeds. Stem cell development might be driven by a combination of intrinsic temporal programs and extracellular signals from the changing environment of the developing brain (Temple 2001).

After removing the mitogenic factors, and in the presence of serum, these cells are capable of differentiating into the three lineages of neural cells: astrocytes, neurons, and oligodendrocytes. They can even give rise to different types of neurons depending on the culture conditions, including dopaminergic neurons if LIF (leukemia inhibitory factor) is added to the medium (Galli et al. 2000b). Surprisingly, recent studies show that the differentiation potential of these cells under certain conditions is wider than previously thought. Sublethally irradiated mice were divided into two groups, and both were injected with genetically labeled bone marrow and neural stem cells, respectively. Grafted cells differentiated into blood cells in both groups (Bjornson et al. 1999). These results have not been reproduced by other groups, which led to this conversion being questioned (Magrassi et al. 2003; Morshead et al. 2002; Yusta-Boyo et al. 2004).

Another experiment consisted of coculturing NSCs and differentiating muscle cells. As a result, skeletal muscle fibers were obtained (Galli et al. 2000a). Similarly, NSCs injected into the embryonic blastula differentiated into a variety of tissues, including renal, cardiac, and hepatic tissues (Clarke et al. 2000).

A few years after this discovery, neurosphere cultures were obtained from different SVZ regions of the human brain (Sanai et al. 2004).

2.3.1
Limitations of the Neurosphere Assay

Our increasing knowledge of stem cell biology has resulted in a re-evaluation of the stem cell concept as it was first established. Stem cells should not be defined in terms of their individual properties in an artificial assay, which itself may alter their characteristics. Stemness is not a property of the cells but a spectrum of capabilities, and the distinction between stem cells and progenitor cells cannot be assessed in vitro.

The clonal nature of a given neurosphere cannot be affirmed, and clusters may fuse at even an ostensibly low density. Even clonal neurospheres consist of a mixed population of progenitor cells and NSCs, and therefore the stemness of a single neurosphere depends on the percentage of NSCs that it contains (Suslov et al. 2002). It has been argued that the multipotentiality exhibited by a neurosphere is due to the presence of the different uni- and bipotential populations in it (Gabay et al. 2003). Thus, neurospheres may represent a more advanced state than bona fide stem cells, and hence a less plastic stage (Fig. 7a).

Neurosphere assays have also been questioned from a molecular viewpoint. Studies on the gene expression profiles of operationally defined NSCs (cells fulfilling the abovementioned properties) compared to SVZ-derived cells revealed striking differences. The profiles of NSCs, hematopoietic stem cells (HSCs) and embryonic stem cells (ESCs) shared a number of genes that might be considered stem-like genes. However, SVZ-derived cells expressed only a few of these genes and displayed a profile that was more consistent with differentiated neural cells. Interestingly, when the operationally defined NSC clone was cultured as a neurosphere, its expression pattern shifted toward a more differentiated pattern, suggesting that the neurosphere, without any other type of validation, may be a poor model for predicting stem cell attributes (Parker et al. 2005).

2.4
Fluorescence-Assisted Cell Sorting (FACS) Analysis in Stem Cell Research

The need for novel strategies that may identify and separate pure populations of NSCs and progenitors has encouraged the employment of FACS analysis in the field of neuroscience (Fig. 7b). Several FACS strategies have recently been applied to sort NSCs from the nervous system using positive and/or negative selections. Uchida and colleagues (2000) showed that a cluster of a differentiation (CD) marker (CD133) was indeed present in hematopoietic cells, and is expressed by a small fraction of fetal brain cells in humans. Positive sorting of the $CD133^+$ population revealed that these cells fulfilled the in vitro stem cell properties while $CD133^-$ cells did not. Just a small fraction of these neurospheres, however, were capable of generating secondary spheres after subcloning.

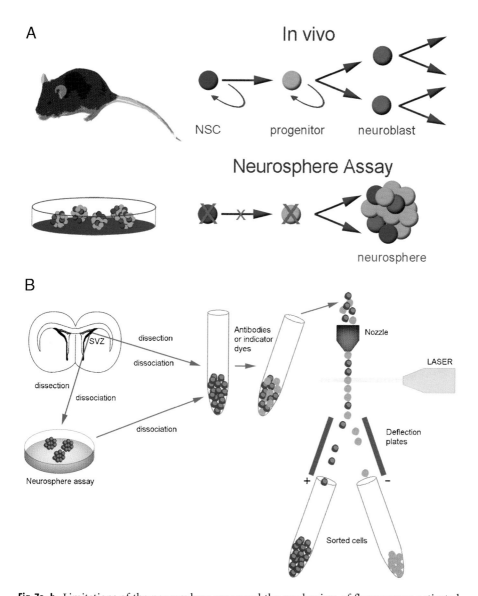

Fig. 7a–b Limitations of the neurosphere assay and the mechanism of fluorescence-activated cell sorting (FACS). **a** In vivo, the lineage progression from stem cell to migratory neuroblasts has been studied and is widely accepted. The in vitro neurosphere assay has been used as a model to study the behavior of stem cells under several conditions, and a parallel between both situations is widely accepted. However, the in vitro assay does not represent a homogeneous population of cloning stem cells, but rather a heterogeneous mixture of cell types in different states of maturity that partially represent *stemness*. **b** Another experimental approach used in NSC research is cell sorting (FACS). Cells are obtained from the tissue sample or culture. After dissociation, single cells are incubated with an indicator dye or specific fluorescent antibodies. The cell sorter counts and separates cells one by one, and sorts labeled and unlabeled cells via laser detection of the dye or fluorescence and by a system of electric charges which allow the purification of the different subtypes and further experimentation

Another FACS strategy involving surface markers, mCD24 and peanut agglutinin (PNA), was applied to adult SVZ in mice (Murayama et al. 2002; Rietze et al. 2001). Cells expressing low levels of both PNA and mCD24 behaved as stem cells do in vitro. A subpopulation of these cells was actually pluripotential, and generated both neural and non-neural phenotypes. Unfortunately, this model presented some limitations, since the quantity of cells obtained from in vivo dissection was minute, and the need for expansion in culture could alter the intrinsic features.

The use of a genetic model combined with a FACS analysis was applied to isolate NSCs from the embryonic and adult CNS. The GFP expression was dependent on nestin expression through a nestin enhancer. Studies revealed that the GFP$^+$ fraction formed neurospheres which could be subcloned and exhibited multipotentiality (Kawaguchi et al. 2001; Roy et al. 2000a). This fraction existed as a larger proportion in the embryonic brain than in the adult brain. The limitation of this model is that nestin is not an exclusive marker of NSCs, so it may be necessary to combine it with the sorting of other surface markers.

Molecular and cellular studies of FACS-sorted NSCs and lineage-restricted progenitors have further demonstrated clear differences in gene expression patterns and functional properties. It is evident that the use of FACS has led to varying degrees of success to date, even though none of the assays has provided a compelling source of a pure population of NSCs. In any case, the reliability of this FACS strategy and its applicability indicate that this method possesses great potential in the future of neurogenesis (Maric and Barker 2004).

2.5
Transgenic Animals and the *cre–lox* System

Mice have been used extensively as a genetic tool to introduce mutations into certain genes and to study the resultant biological effects. Transgenic mice expressing visible markers under specific promoters have enabled us to follow the cell progeny of NSCs or to investigate the origin of radial glia; such visible markers include GFP (green fluorescent protein) (Fig. 4b), AP (alkaline phosphatase), or LacZ (beta-galactosidase) (Fig. 4e,f). GFP is a fluorescent protein that allows for confocal analysis and the study of colocalization with other markers, three-dimensional structures, and cell morphology. However, it is not quite as useful for EM because sample processing is complex and deteriorates the tissue (it consists of DAB immunostaining). Gold immunostaining damages the tissue too, but the penetration of the antibodies is lower. AP is not applicable to EM due to technical reasons. LacZ is excellent for EM but only permits the cell body to be studied. It deposits perinuclearly and in the endoplasmic reticulum, but not in cellular expansions.

Lately this approach has undergone a revolution due to the application of the *cre–lox* system. This system consists of a DNA integrase, the *cre* recombinase from the P1 phage (a specific bacterial virus), and a coupled nucleotide sequence, the *loxP* site, such as the *cre* DNA substrate. *loxP* is a sequence of 34 base pairs (bp) containing two 13 bp inverted repeats flanking an 8 bp core sequence. Two *cre* molecules bind to each *loxP* site, one on each half of the palindrome (Van Duyne 2001). *Cre* molecules bind to DNA and then form a tetrameric complex to bring two *loxP* sites into proximity, thus creating the orientation that is essential for the resulting recombination. The placement of recombination sites in the genome and the subsequent targeted expression of the recombinase have allowed the development of genetic switches that can either ablate or turn on any desirable genes in transgenic or gene-modified mice (Sauer 1998).

In the study of neurogenesis, this system enables the detection of specific cell populations or the more active creation of conditional mutants to study the functions of target genes. The first case, for example, requires the use of transgenic mice with a gene construction under a general promoter, such as beta-actin and a loxP/loxP-flanked stop sequence. When the virus that expresses the cre recombinase is injected into the brains of these transgenic mice, the stop sequence is removed from infected cells and the marker starts to constitutively express, so that the cells can be visualized (Fig. 8a). This type of experiment was used to demonstrate that radial glia in the neonatal brain were the source of neurons, as were the glia in the subventricular zone and olfactory bulb (Merkle et al. 2004). This can also be achieved by using two transgenic animals where one animal constitutively expresses cre recombinase while the other expresses the target gene flanked by two loxP sites (Fig. 8a). Furthermore, gene construction might correspond to a visible marker (Fig. 8b).

The creation of conditional mutants depends on the targeted deletion of certain genes that are flanked by *flox/flox*, a subtype of the *cre–lox* system sites in transgenic animals. The great advantage of this relatively recent tool is that it allows certain genes in the adult animal to be conditionally knocked out, thus avoiding the compensatory effects observed in developmental mutants.
Besides all of the techniques explained above, other tools such as cell transplantation and electrophysiology allow us to follow differentiation and integration into the nervous tissue in vivo.

2.6
Transplantation of Adult NSCs

Adult NSC transplantation is a new approach to stem cell-based therapies that may contribute to the regeneration of damaged areas after brain injury. It has been implemented over the last decade through the discovery of new cell-labeling procedures that allow the fate of grafts in the target tissue to be followed. New advances in surgery and stereotaxis have also allowed positive results to be obtained when transplanting cells or grafts into the brain.

The source of cells for transplantation can vary from SVZ-derived adult NSCs to bone marrow stem cells, carotid body, or embryonic stem cells. Once obtained, cells can be amplified and introduced into the brain by stereotaxic procedures or

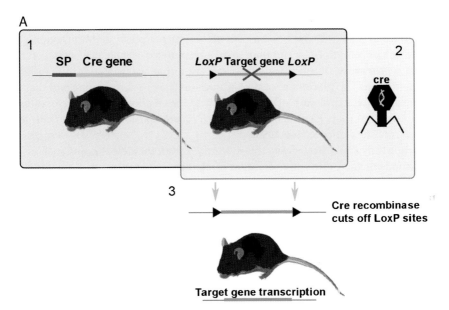

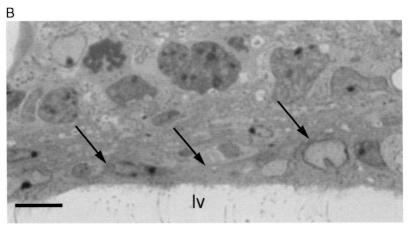

Fig. 8a–b Transgenic animals used in NSC research and the *cre–lox* system. **a** The *cre–lox* system can be applied by crossing two transgenic mice, one expressing the *cre* recombinase gene under a specific promoter (*1*) with another mouse which contains the target gene flanked by the *loxP* sites (overlapping the region between *1* and *2*), stopping its expression. Their progeny will have expressed the gene in those cell types that express the specific promoter. Thus the *cre* recombinase will remove the *loxP* sites and allow the gene to be transcripted (*3*). One alternative, among others, is to transfect a transgenic mouse carrying the *loxP* sites (overlapping the region between *1* and *2*) with a virus containing the *cre* recombinase in its genetic program (*2*). The effectively infected cells will express the target gene (*3*). **b** A semithin section of the adult mouse SVZ with beta-gal staining. This picture corresponds to a transgenic animal with the construct noggin–lacZ. Beta-gal staining only labels those cells that express noggin, which, in this case, are ependymal cells (*arrows*), *scale bar* 10 µm (*SP*, specific promoter)

by endovenous injection. In that respect, the systemic injection of cells has demonstrated in several assays on stroke or neurodegeneration (in injured regions) that the endovenously administered cells can reach certain areas of the brain (Pluchino et al. 2003; Politi et al. 2007).

Multiple strategies have been used to label the cells to be transplanted, such as male-to-female transplantation, transgenic animals (as mentioned in this chapter), nucleotide analogs (BrdU, 3HT), or iron particles, among others. Although easy to use, extrinsic markers have drawbacks. For instance, if grafted, cells may die. The marker diffuses and can be taken up by NSCs. Rather than the grafted cells, the labeled population observed may represent those already present in the animal (Burns et al. 2006). The use of viruses entails the problem of silencing expression after transplantation. Therefore, these labeling techniques must be evaluated critically, and simultaneous protocols may be used to confirm the results.

2.7
Integration and Functionality of Newborn Cells in the Adult Brain

To evaluate the process of neurogenesis, it is necessary to study the proliferation, migration, and differentiation of newborn cells, and it is also essential to quantify the percentage of these cells that finally integrate into the target tissue.

Such quantification requires the use of stereology. Stereological techniques sample plane sections (two-dimensional) from an object and—by combining geometric probability and statistics—extrapolate them to the three-dimensional material. Many scientific errors arising from the misinterpretation of plane sections have been overcome.

To establish whether a newborn neuron is functional, connectivity with other cells must be demonstrated, as well as the electric activity that the cell needs to play its role in the target tissue. An elegant work (Carleton et al. 2003) that uses electrophysiology combined with the transgenic labeling of animals has shown that the neurons that are newly incorporated into the OB (and originated in the SVZ) are capable of integrating into the pre-existent circuitry.

Behavioral neuroscience is also fundamental to the study of neurogenesis. This allows the formation and integration of new neurons to be correlated with the recovery of lost functions after brain injury, which is secondary to ischemia or neurodegenerative diseases. However, its complexity precludes us from including behavioral methods of studying neurogenesis in our study.**Integration and Functionality of Newborn Cells in the Adult Brain**

3
Neurogenesis in the Intact Adult Mammalian Central Nervous System

3.1
Description of Neurogenic Regions in the Adult Mammalian Brain

There are two well-accepted neurogenic regions in the adult mammalian brain: the olfactory bulb and the hippocampus (Fig. 9). However, stem cells with neurogenic potential have also been found in the third ventricle, in the central canal of the spinal cord, and more recently, in the subcallosal region (SCR), a region between the corpus callosum and the hippocampus.

3.1.1
Subventricular Zone

The main neurogenic region in the adult mammalian brain is the olfactory bulb (OB), where new neurons (granular and periglomerular subtypes) are formed de novo. The majority of these new neurons are generated in an area that underlies these lateral ventricles, the subventricular zone (SVZ). After being produced, a large percentage of these cells migrate toward the OB following a pathway (5–6 mm in the rodent) called the rostral migratory stream (RMS).

The first works dedicated to neurogenesis, performed in the middle of the twentieth century, were based on morphological observations made via light microscopy. They reported that, aside from the multiciliated cells that cover the ventricle, two cell types exist: light and dark cells. However, since then, the combination of electron microscopy (EM) studies and three-dimensional reconstructions with the use of cellular markers has led to a precise description of the rodent SVZ as a first step in the identification of adult NSCs in the adult brain.

3.1.1.1
Characterization of SVZ Cell Types with Light Microscopy

The SVZ of the adult mouse brain is a discontinuous region of cells next to the ependymal lining. This SVZ is most evident, and involves up to four cell layers, in the lateral wall of the lateral ventricle facing the striatum (Mitro and Palkovits

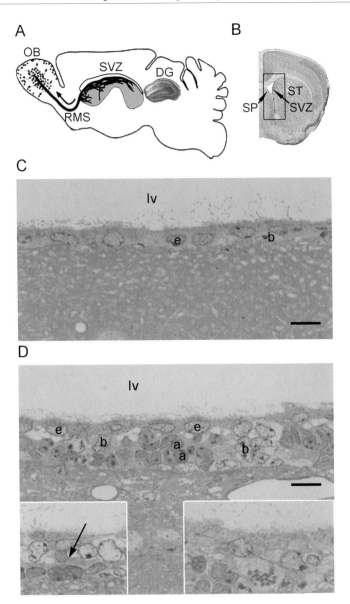

1981) (Fig. 9). The medial wall facing the septum is largely devoid of cells—comprising up to one cell layer—with the exception of the most anterior part of the anterior horn, where dark and light cells, which are similar to those in the lateral wall, are found. The roof of the lateral ventricle is almost devoid of SVZ cells. The same cannot be said of the septum, as we have found only light cells (astrocytes) and ependymal cells. Ependymal cells are flat in the septum and astrocytes form a thin band underlying the ependymal layer with protruding nuclei. In the anterior levels of the rostrocaudal axis, cells constituting the SVZ occasionally expand toward the ventral portion of the septum.

The SVZ is formed from several layers of cells (usually from two to five). However, the number of cell layers varies depending on the animals, ages, and the different rostrocaudal levels. At the posterior and medium levels, the SVZ comprises only two layers of cells, resembling the septum, and consists of only astrocytes and ependymal cells. The maximum cell density of the SVZ is found in the dorsal and ventral regions of the anterior levels.

Using light microscopy, SVZ cell populations can be identified (even without being able to ensure 100% reliability until observed under EM) with the following features:

(a) Type A cells or immature neurons look small and dark, and form clusters of cells close to the ventricle. These cells are spindle-shaped and form clusters, but they may appear to be round depending on the histological section. The nucleus is fusiform or round, while chromatin is dense but dispersed with two to four nucleoli. Cytosol is scarce but densely packed.
(b) Type B cells are large and irregular with light and invaginated nuclei. Cytosol can be dark or light depending on the amount of intermediate filaments, but is always less dark than Type A cells. These cells are frequently in contact with ependymal cells. They can form small clusters in the boundaries of the clusters of A cells.

Fig. 9a–d Proliferative regions in the adult mouse brain. Subventricular zone (SVZ) under light microscopy. **a** This diagram shows a sagittal view of the mouse brain where the main proliferative regions (SVZ and dentate gyrus—DG—of the hippocampus) and neurogenic (olfactory bulb—OB, and DG) are represented. The rostral migratory stream (RMS) is the pathway that newly formed neurons from the SVZ take to reach the OB after migrating several millimeters of the distance. The stem cells in the DG give rise to neurons that migrate much less and finally differentiate within the hippocampus. **b** This picture shows the ventricle in a coronal section. The medial wall corresponds to the septum (SP) (**c**), while the lateral wall constitutes the SVZ (**d**). **c** The septum does not host neural stem cells and is poorly populated with astrocytes and ependymal cells, *scale bar* 10 μm. **d** In contrast, the SVZ is composed of different subpopulations of cells (described in the text). In the *subset on the left*, a Type C cell is depicted (*arrow*); an amplifying precursor. A mitotic figure can be seen in the *subset on the right*, a very frequent finding in the SVZ (*arrow*), *scale bar* 10 μm (*OB*, olfactory bulb; *RMS*, rostral migratory stream; *SVZ*, subventricular zone; *DG*, dentate gyrus; *SP*, septum; *ST*, striatum)

(c) Type C cells are found in the close vicinity of Type A cells (subset to the left of Fig. 9). They share features with subtypes B and A, and share an intermediate phenotype, even when they are the largest cells in the SVZ. It is common to find C cells in the mitosis (subset to the right of Fig. 9). Their nuclei are dark and chromatin is dispersed. Occasionally, clumped chromatin relates to a pre- or postmitotic state. This cell type is the most difficult to identify with LM and EM because it is an intermediate stage between B and A cells.

(d) Type E cells are cubic or columnar cells that line the cavity of the lateral ventricle. They possess cilia in their apical pole and lipid drops in their cytosol. Nuclei are round or oval, and abundant mitochondria, which look like basophilic dots with LM, are dispersed in the cytosol. It is possible to identify neurons and microglial cells in the SVZ with LM.

3.1.1.2
Characterization of SVZ Cell Types with Electron Microscopy

As observed with LM, four distinct types of cells constitute the periventricular SVZ region of the lateral ventricles: (1) Type A cells (migratory neuroblasts), (2) Type B cells (astrocyte-like), (3) Type C cells (precursors), and (4) Type E cells (ependymal cells) (Fig. 10). Refer to Table 3.1 for ultrastructural features of different cell types.

Type A cells are fusiform with one or two processes: dark under LM and electron-dense under EM. The nucleus is dark, occasionally invaginated, and chromatin is mainly euchromatin with two to four small nucleoli. The cytosol is scant, with a large number of free ribosomes, a few short cisternae of a rough endoplasmic reticulum (RER), a small Golgi apparatus, and many microtubules oriented along the axis of the cells. The plasma membrane displays characteristic union complexes with neighboring cells (Fig. 10). PSA–NCAM, Tuj1, and Dcx are useful specific markers. Although not specifically, they also express Dlx2, a transcription factor expressed by precursor C cells in the SVZ. Type A cells have the ability to migrate tangentially along the SVZ and the RMS, where they form chain-like structures (Fig. 11). Three-dimensional reconstructions have shown that all Type A cells contact with astrocytes (Fig. 12). Migratory neuroblasts are not completely surrounded by other

Fig. 10a–e Panoramic view of the adult mouse SVZ and the ultrastructure of Type A cells. **a** A panoramic view of the SVZ under electron microscopy (EM) is shown. The lateral ventricle (lv) is surrounded by the ependymal layer of ependymal cells (e), and underneath, in contact with the ependyma, we see the SVZ, which is composed of different cell types—Types A (**a**), B (**b**) and C (**c**)—with the features described in the text, *scale bar* 5 µm. **b–e** Ultrastructure of a Type A cell. **b** The abundance of free ribosomes is very characteristic of Type A cells, *scale bar* 250 nm. **c** Type A cells have small dictyosomes in their Golgi with short saculae (*asterisk*), *scale bar* 500 nm. **d** Transversal sections of microtubules in the cell body of Type A cells (*long arrow*), and typical cell junctions are also frequent (*short arrows*). Intercellular gaps allow for cell migration (*asterisk*), *scale bar* 200 nm. **e** Centriole immersed in the cytosol directing migration (*arrow*), *scale bar* 200 nm

Subventricular Zone

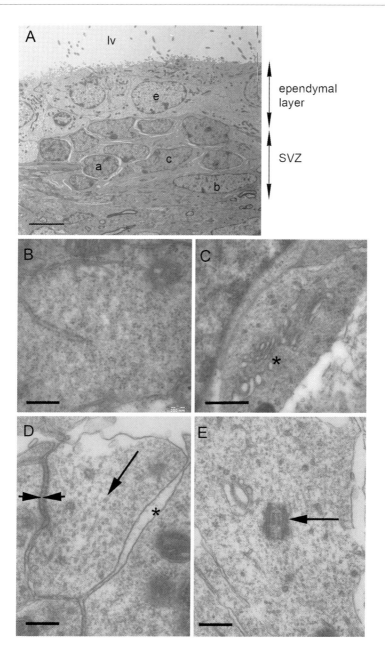

Table 3.1 Ultrastructural features of SVZ cell populations in rodent brain

	Type A	Type B	Type C	Type E
Position of the cell body	Basal	Apical–basal	Large	Apical
	No contact with the ventricle	Occasionally contact the ventricle		Always contact the ventricle
	Elongated shape	Pseudostratified irregular shape		Cuboidal shape
Expansions	Tangential	Radial electron-dense	Not frequent	Radial electron-lucent
		Multiple expansions interspersed between cells		
Nucleus	Round–fusiform	Oval irregular	Highly invaginated	Round–oval
	Dark			
Chromatin	Dispersed with clumped chromatin	Often dispersed		Small clumps of chromatin
Nucleoli	1–2 large	1–2 large		1–2 small
Cytoplasm	Scant and dark	Light	Intermediate features between Type A and B cells	Light in the basal region and dark in the apical region
REL	+	+++		+
Golgi apparatus	Small	Large (thin dictyosomes)		Small (dilated dictyosomes)
Anullate lamellae	No	Yes		Yes
Golgi-associated vesicles	++	++++		+
Free ribosomes	++++	++	++	++
Mitochondria	+	+++ (small and round)		+++ (large and elongated)
Centriole	Close to the nucleus and far from the ventricle	In the apical cytosol		None detected

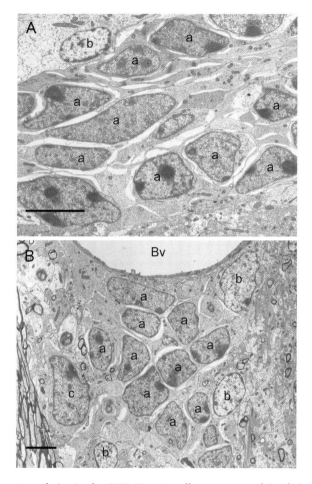

Fig. 11a–b Migratory chain. In the SVZ, Type A cells are arranged in chains surrounded by astrocyte-like cells (Type B cells), and they migrate along the rostral migratory stream toward the OB. **a** A longitudinal section of the chain is depicted, *scale bar* 5 μm. **b** A transversal section of the chains migrating in close relationship with blood vessels (Bv). Note the characteristic spaces between the cells, *scale bar* 5 μm

migratory cells. Type A cells possess a short primary cilium orientated in the direction of migration with a centriole at its base. Abundant microtubules arise from the centriole. This is very interesting because the orientation of the primary cilium and the centriole seem to be involved in directing cell migration. It has been recently demonstrated that the migration of neuroblasts through the SVZ toward the RMS parallels the cerebrospinal fluid (CSF) flow, suggesting a potential role of ependymal cells in their guidance (Sawamoto et al. 2006), probably through repulsive factors such as *Slit2* (and other molecules), secreted by cells from the choroidal plexus and distributed by ependymal cilia movement.

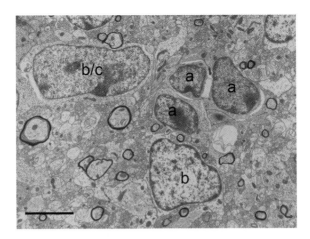

Fig. 12 Type B cell. Type B cells (*b*) accompany migratory cells on their way to the OB. These cells clearly differ to Type A cells (*a*). Their cytosol is lighter, their nuclei are invaginated and they possess intermediate filaments that can be observed at high magnification under EM. In the early RMS, Type C cells can be observed. In the picture, a Type B cell is transforming into a Type C cell (*b/c*). It is commonplace to find cells in intermediate differentiation states along the length of the SVZ, *scale bar* 5 µm

Type B cells have irregular contours that fill the spaces between neighboring cells profusely. Cytoplasm is lighter, with a few ribosomes, and it is enriched with intermediate filaments formed from the GFAP protein, among others, which are highly expressed by mature astrocytes. Other intermediate filaments present in B cells are vimentin and nestin. These cells have a nucleus that is deeply invaginated with dense chromatin. B-cell astrocytes establish cell-to-cell contact with other astrocytes and ependymal cells through GAP junctions, but never with Type C cells or migratory neuroblasts. These cells constitute a dense network underlying the ependymal layer and they surround the chains of A cells (Fig. 12).

Occasionally, B cells contact the lumen of the ventricle and exhibit a short straight single cilium (Fig. 13). This cilium has a 9 + 0 organization. It can be found to exist partially inside the cell, and it displays a structure that is very similar to the cilia found in radial cells in reptiles and birds. Two recent publications by our group show the importance of cell signaling through the primary cilia in cerebellum (Spassky et al. 2008) and dentate gyrus (Han et al. 2008), which are mandatory for adult hippocampal neurogenesis. It has been suggested that it may play an important role in proliferation (Singla and Reiter 2006) and development (Davis et al. 2006). In addition, radial cells in the mammalian embryo also display a single short cilium. The number of B cells contacting the ventricle increases after injection of EGF (Doetsch et al. 2002) or ephrins (Conover et al. 2000), or after any process that implies regeneration of the SVZ. This phenomenon remains unclear, but it may be related to the uptake of the factors in the CSF that regulate cell proliferation. We also know that the number of B cells that contact the ventricles in neonates—where neurogenesis is more prominent—is very high, while it is low in adults, except in the abovementioned situations.

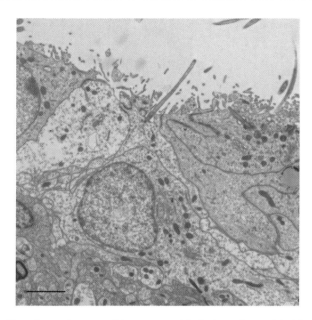

Fig. 13 Single cilium from a Type B cell in contact with the lateral ventricle. The basal body of the cilium can be occasionally observed within the cell. This cilium can originate from either the soma or a cytosolic expansion, *scale bar* 2 μm

In mice, two types of B cells were described, Type B1 and B2 (Fig. 14). Type B1 astrocytes are darker, have more organelles in their cytoplasm, and are the cells that send the primary cilium that contacts the lumen of the ventricle. Therefore, B1 cells resemble radial glia and are in close contact with the ependymal cells in the SVZ. Chromatin is relatively dispersed in the nucleus. B2 astrocytes are smaller, and the primary cilium directs to the underlying parenchyma. These cells resemble those astrocytes in the RMS. Type B2 astrocytes have clumped chromatin, a scarce cytosol with few organelles, and they tend to localize at the interface with the striatal parenchyma. Ultrastructurally, these cells are phenotypically identical to the astroglia found in the striatum or cerebral cortex. Intermediate filaments in these cells are found in the expansions, far from the cell body. In the septum, there are only B2 astrocytes, which constitute a thin monolayer that separates ependymal cells from the underlying neuropile. Even though attempts have been made to classify SVZ astrocytes based on their morphology, it is not clear (i) which are the bona fide stem cells, or (ii) whether the different morphologies described for each subtype correspond to different states of maturity. Neither is there enough literature to support any conclusion. Even though we know that there are astrocytes that come into contact with the lumen of the ventricle, the question is: is this a static or a dynamic event? There is no molecular evidence of a marker that allows us to classify the astrocytes. What we do know is that B2 cells do not come into contact with the ventricle and they ultrastructurally form glial tubes, which does not rule out the possibility of them acting as NSCs. In fact, some authors were able to obtain clones of neurospheres from cells derived from the RMS even when it was unclear whether these cells were astrocytes (Gritti et al. 2002).

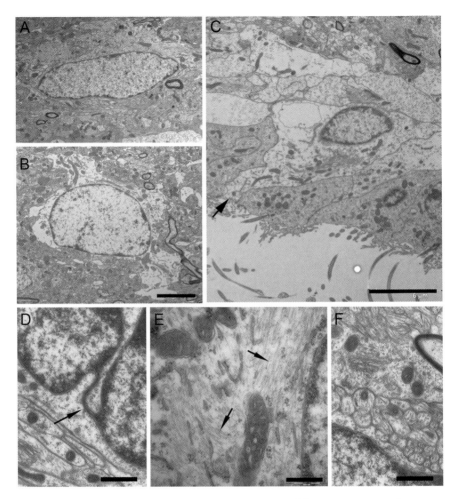

Fig. 14a–f Types of Type B cell. **a, b** Within the SVZ, we can delineate two types of astrocytes based on morphologic criteria, Type B1 (**a**) and Type B2 (**b**), *scale bar* 2 µm. B1 astrocytes are rich in intermediate filaments and their cytosol is very electron-dense compared to B2, as seen in the panoramic views. **c** B1 astrocytes contact the lumen of the ventricle (*arrow*), *scale bar* 5 µm. **d** Both types of astrocytes have invaginated nuclei typically with thin structures (*arrow*), *scale bar* 50 nm. **e** The Type B1 cell contains abundant organelles such as RER, Golgi, mitochondria, and intermediate filaments (*arrows*) are abundant in the B1 astrocyte, mostly in the soma, *scale bar* 10 nm. **f** It is very characteristic that Type B1 astrocytes ensheathe migratory cells and come into contact with myelinic and amyelinic axons, *scale bar* 50 nm

Type C cells (transit-amplifying precursors) share some features with Type A cells, but they are more electron-lucent, have fewer ribosomes, their nuclei are larger, spherical and frequently invaginated, and they are not found in the RMS (Fig. 15). They have several large reticulated and atypical nucleoli. Their cytoplasm

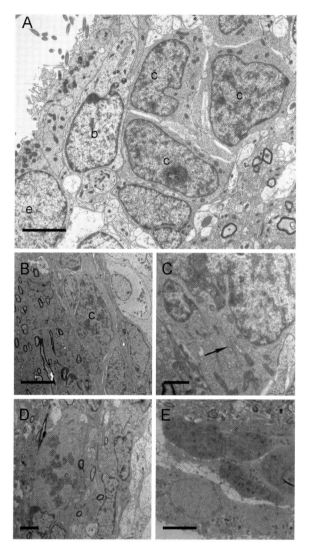

Fig. 15a–e Type C cell. **a** Cluster of C cells immersed within the SVZ, *scale bar* 2 mm. **b** Panoramic view of a C cell in the SVZ showing a very large and invaginated nucleus with large nucleoli. In the SVZ, the Type C cell associates with Type A cells and is barely found in the RMS, *scale bar* 1 mm. **c** Detail of the cytosol. These cells contain numerous organelles with large dictyosomes (*arrow*) and an important RER with ribosomes, even more dense than in astrocytes or B cells. However, they have fewer ribosomes and microtubules compared to the migratory neurons that come from C cells, *scale bar* 1 mm. **d** A mitotic figure corresponding to a C cell—a frequent finding in the SVZ, as these cells constitute the fast-proliferating precursors, *scale bar* 2 mm. **e** Dlx2 pre-embedding immunostaining specific for C- and A-cell populations, *scale bar* 5 mm

contains a large Golgi apparatus, and they do not possess bundles of intermediate filaments or microtubules. C cells form clusters in the proximity of the chains formed by migratory cells, and they occasionally establish small junctional complexes with them. Even though these cells proliferate at the highest rate in the SVZ and they express the transcription factor Dlx2, they are negative for markers of immature neurons such as PSA–NCAM. An EM analysis does not identify around 20% of Type C cells. This is the consequence of cells holding intermediate characteristics between different cell types. Cells with intermediate filaments and abundant ribosomes will correspond to B–C transitional cells, while those cells with scarce microtubules and ribosomes will correspond to C–A cells.

Type E cells (ependymal cells) are large and cubical, and they form a monolayer that separates the lateral ventricle from the surrounding parenchyma. One main feature of these cells is the presence of cilia and numerous microvilli in their apical pole (Fig. 16 and Fig. 17). Lateral processes are interspersed between contiguous cells and remain in close contact through many junctional complexes, such as tight junctions and desmosomes. Recently *Numb*, a new protein, was reported, which is involved in the cohesion of ependymal cells. When *Numb* is absent, Type E cells detach themselves from neighboring cells, and the SVZ invades the gap toward the ventricle (Kuo et al. 2006). Interestingly, these cells proliferate and behave like Type C cells, giving rise to small tumor formations. Ependymal cells, however, do not react, and they are never replenished. Nonetheless, astrocytes constitute layers that line the surface, isolating the neuropile from the CSF.

The cytosol of Type E cells is electron-lucent with few organelles, except for the basal corpuscle of the cilia where a large number of mitochondria accumulate. The majority of these mitochondria are bound to the base of the cilia through filamentous structures. The nucleus is spherically composed of sparse chromatin with clusters of heterochromatin associated with the nuclear membrane. In the basal pole, some cells send an expansion that interacts with the basal lamina of blood vessels. Such expansions are frequent in dorsal and ventral horns, and these radial expansions can reach considerable lengths after a stroke or brain injury. A relatively small number of intermediate filaments can be observed in the ependymal cells of rodents compared to upper mammals (humans or primates), where they are abundant. Immunohistochemically, these cells express vimentin, S-100, and CD-24.

Fig. 16a–d Ependymal cells. **a** Typical ependymal cells in contact with the ventricle with abundant cilia and microvellosities. Mitochondria tend to concentrate in the apical pole of the cell. In the upper right-hand corner, there is a subset with detail of a basal body of the cilium, *scale bar* 5 μm. **b** Lipid drops (*asterisk*) are a frequent finding in the ependymal cells. Observe the increased density of the cytosol in ependymal cells compared to astrocytes, *scale bar* 1 μm. **c** In this picture, the arrows point to two different types of cell junctions, union adherens (*white arrows*), tight junctions (*black arrows*), and desmosomes. In the cytosol, intermediate filaments are abundant, *scale bar* 200 nm. **d** Scanning electron microscopy allows us to study the cell surface. Multiple cilia and microvellosities are characteristic of ependymal cells, *scale bar* 2.5 μm

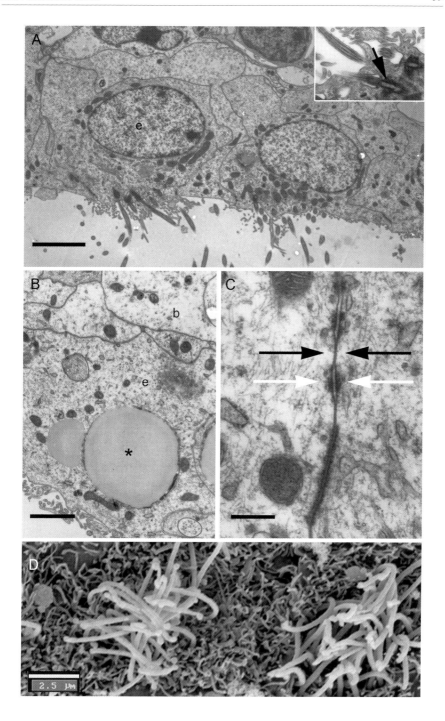

Fig. 17 Surface of the lateral ventricle. Through their cilia, ependymal cells play an important role in CSF flow, transporting molecules toward the brain parenchyma, and recent studies demonstrate a relationship with cell migration. This picture corresponds to a scanning micrograph of the ependymal wall of the lateral ventricle, *scale bar* 10 μm

Other cell types may be found in smaller percentages in the SVZ. Some neurons, pyknotic cells, oligodendrocytes and microglia were sporadically found in the SVZ, mixed with the populations described above (Fig. 18).

3.1.1.3
Rostral Migratory Stream

The SVZ is continuous with the RMS, which is a pathway that rostrally expands toward the OB. It is composed of Type A cells that are surrounded by Type B cells to form a sheath in order to isolate the future neurons from the brain parenchyma, thus constituting a type of structure that is known as a glial tube (Jankovski and Sotelo 1996; Lois et al. 1996; Peretto et al. 1998). Astrocytes remain fixed while A cells move along these structures. This area can easily be identified with LM, as many blood vessels are found in the region, since they accompany cell chains. These cells are likely to have a nutritional function.

The migration of new neurons is favored by the cell-to-cell interaction that takes place between Type A cells and the astrocytes from the glial tubes (Fig. 19). Additionally, ependymal cells that allow the movement of the CSF by synchronized and polarized cilia movement play a major role in migration, as they direct the fluid within the ventricles. The direction of the CSF flux coincides—and not incidentally so—with the direction of the migratory cells on their way toward the OB. The CSF is produced by the choroidal plexus located in the ventricular system. Another molecule, a repulsive protein produced by the same cells, *Slit2*, seems to favor migration (Sawamoto et al. 2006). The movement of the cilium can be blocked by using conditional mutant animals or an in vitro infusion of *Slit2*, and it is capable

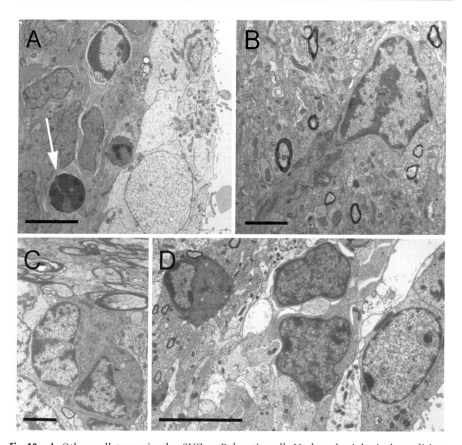

Fig. 18a–d Other cell types in the SVZ. **a** Pyknotic cell. Under physiological conditions, pyknotic cells are occasionally found within the SVZ (*white arrow*). Their ultrastructure does not differ from the classical descriptions of pyknotic cells during development. They possess a compacted nucleus with highly condensed chromatin. It is usual to find microglial cells in the close vicinity. The number of pyknotic cells increases greatly after irradiation, mutagenic insult or antimitotic treatment, *scale bar* 5 µm. **b** Microglia. Microglial cells are distributed homogeneously along the SVZ. They are star-shaped with thin expansions interspersed between other cell populations. These cells are highly electron-dense, and present abundant and large lysosomes within their cytosol. Under pathologic conditions, in which cell death increases, microglial cells increase in parallel and phagocyte cellular remains, *scale bar* 2 µm. **c, d** Oligodendrocyte. It is not very commonplace to observe mature oligodendrocytes in the SVZ. These cells are always associated with bundles of myelinated axons in the striatum, unlike the dorsal regions close to the corpus callosum, where the number of oligodendrocytes increases. Ultrastructurally, they are small dark cells characterized by a spherical nucleus with clumped chromatin attached to the nuclear membrane. Within the cytosol, the cisternae of RER are short and dilated. **c** *Scale bar* 1 µm, **d** *scale bar* 5 µm

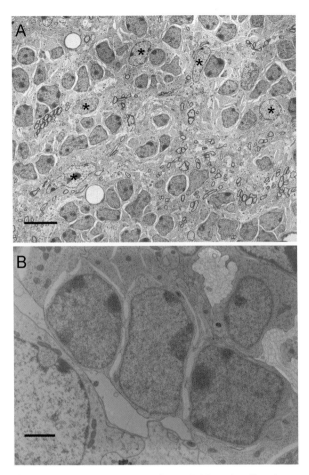

Fig. 19a–b Rostral migratory stream. **a** From the SVZ, Type A cells migrate toward the OB by means of tangential migration. These cells form multiple chains that are surrounded by and totally isolated from the remaining parenchyma through astrocytes (*asterisk*). These structures constitute the RMS, which is about 3–5 mm long and ends in the OB, *scale bar* 10 μm. **b** Migratory cells are located in the center of the chain (*shaded in red*), while astrocytes surround them (*shaded in blue*). Both cell types are in contact with amyelinic axons (*shaded in yellow*) on their way toward the OB, *scale bar* 2 μm

of disorganizing the structure of the migratory chains in the RMS. Although this is a tremendous step forward in explaining tangential migration, not all of the cues involved in this process have been described entirely and/or are understood yet.

3.1.2
Hippocampal Dentate Gyrus

The other region considered to be neurogenic in adult mammals is the dentate gyrus (DG) of the hippocampus. It is thought that hippocampal neurogenesis is related to learning and memory (Gould et al. 1999c; Shors et al. 2001). The DG consists of a broad band of neurons which group into two branches. The cytoarchitecture is simple. The granule cell layer (GCL) is formed exclusively from granule neurons, a few microglial cells and some astrocytes interspersed between the neuronal bodies. Deep in this region there is a thin layer where progenitors reside: the subgranular zone (SGZ) (Fig. 20). In semithin sections, small clusters consisting of light and dark cells with blood vessels in their close vicinity are frequently observed. These clusters or neurogenic niches are homogeneously distributed.

The neurogenic niche in the SGZ is unique in that, unlike the SVZ or the VZ of adult birds (Alvarez-Buylla and Nottebohm 1988), it is separated from the walls of the ventricles or the ependymal layer (Alvarez-Buylla and Lim 2004). SGZ cells express GFAP and share the characteristics of astrocytes. Two types of astrocytes have been identified in accordance with morphological criteria: horizontal and

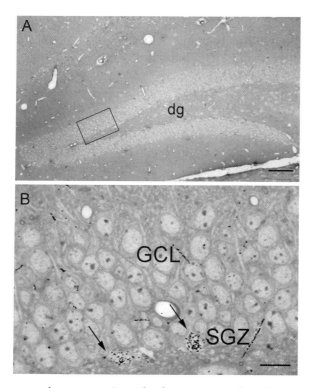

Fig. 20a–b Hippocampal neurogenesis. **a** The dentate gyrus of the hippocampus (dg) is a region where adult neurogenesis has been widely accepted, *scale bar* 50 µm. **b** Subset of the dentate gyrus (*square* in **a**). Neural stem cells reside in the subgranular zone (SGZ). These cells incorporate 3H-thymidine because some of them are in the S-phase, preparing to divide. Their progeny migrate to the granular cell layer (GCL) and give rise to more mature cell types with neuronal phenotypes, *scale bar* 10 µm (*dg*, dentate gyrus; *GCL*, granular cell layer; *SGZ*, subgranular zone)

radial astrocytes. Radial astrocytes have a large cell body with a major radial process that penetrates the GCL and crosses over the GCL. These expansions constitute thin basal lamellae, are oriented tangentially along the SGZ, and send thin lateral processes intercalated extensively between granule neurons that isolate them from the neuropile. The radial process also branches profusely in the molecular layer (ML). Some authors consider the radial expansion to be a scaffold for the migration of newly formed neurons. In contrast, horizontal astrocytes have no radial projection but extend, instead, to the branched processes that are parallel to the SGZ and the thin short secondary branches into the hilus and the GCL. Horizontal astrocytes are generally elongated, whereas radial astrocytes have a round, polygonal, or triangular cell body. Under EM, these cells are characterized by a light cytoplasm containing a few ribosomes, scarce RER, a dense network of intermediate filaments, a thin Golgi apparatus, and irregular contours with a plasma membrane and processes in contact with neighboring cells. The nuclei of these cells generally contain sparse chromatin in granules, but variable heterochromatin aggregation suggests that these cells may differ in proliferation stages. Radial astrocytes have more organelles, polyribosomes, and lighter mitochondria compared with horizontal astrocytes. In addition, the large bundles of intermediate filaments present in the main process of radial astrocytes are less prominent in horizontal astrocytes. Radial astrocytes are nestin$^+$, whereas horizontal astrocytes stain with antibodies against S-100, a calcium-binding protein that is specifically expressed in some astrocytes and ependymal cells. Glutamine synthetase, an enzyme that converts glutamate into glutamine, was expressed in some but not all the horizontal astrocytes.

Both types of astrocyte have been identified as the source of new neurons in the adult hippocampus (Seri et al. 2001). These cells are responsible for the formation of new granular neurons that migrate to a lesser extent than the cells in the RMS, and which reach their final destination in the GCL. Given their similarities with the B cells found in the SVZ, these cells have been named SGZ B cells. The complex morphology of SGZ astrocytes with multiple processes that penetrate the GCL (Cameron et al. 1993; Eckenhoff and Rakic 1991; Kosaka and Hama 1986) goes against the notion that neural progenitor cells are undifferentiated with no role to play other than the generation of new neurons. Thus, SGZ astrocytes seem to combine the functions of both progenitors and glial cells, since their radial processes that intercalate among mature granule neurons could potentially carry proliferation, differentiation, or maturation information to the newly formed neurons.

A different type of cell with a dark nucleus and GFAP immunoreactivity was found in the SGZ. These cells probably correspond to the small, dark-staining progenitors that others observed previously (Altman and Das 1965; Cameron et al. 1993; Kaplan and Bell 1984). These cells specifically express Dcx and PSA-NCAM, and were named Type D cells. Type D cells do not resemble any type of cell in the SVZ, and clearly differ from astrocytes. They are labeled in 3HT experiments for short survival times (one week). These cells are derived from astrocytes and probably function as transient precursors in the formation of new neurons. Ultrastructurally speaking, they have smooth contours, a dark, scant cytoplasm with many polyribosomes (more than astrocytes but less than granule neurons), and darker nuclei. Their mitochondria are lighter than those of astrocytes, and the endoplasmic reticulum is larger than in astrocytes but smaller than mature granule neurons. D cells generally form clusters of 2–4 cells. They are partially isolated from the environment by expansions of radial glia, forming basket-like structures that might be considered radial proliferative units (Seri et al. 2001) (Fig. 21). An ultrastructural gradient is observed from D cells, which look more like astrocytes and cells that closely resemble granule neurons. By using EM and confocal microscopy combined with PSA–NCAM immunohistochemistry, three subtypes of D cells have been characterized. D1 (48%) cells are small, have little cytoplasm, and no processes or very thin protrusions that are usually in the plane of the SGZ. D2 (32%) cells have a short, thick process that sometimes bifurcates. More subdivisions have been made in this group based on the orientation of the process. D3 (20%) cells have the characteristics of immature granule neurons; a prominent, frequently branched, radial process that extends through the GCL toward the ML, and thin processes projecting into the hilus (most likely corresponding to the axon). Their somas are round to polygonal. Occasionally these cells have other branches that interdigitate between mature granule neurons. D3 cells are generally found in the interface of the GCL and the SGZ but, occasionally, the cell bodies of D3 cells can be found deeper in the GCL.

During development, at embryonic day E18, the rodent dentate gyrus receives progenitor cells from the neuroepithelium of the ventricular wall in front of the hippocampus, the hippocampal SVZ (Pleasure et al. 2000). This cell migration even extends postnatally, and the progenitors establish a reservoir of cells in the adult DG that can differentiate in both neurons and glia. Some authors believe that a subpopulation of this reservoir may be of the astrocytic lineage, and may constitute the neurogenic niche in the adult SGZ that we described previously as adult hippocampal stem cells (Navarro-Quiroga et al. 2006).

3.1.3
Concept of Neurogenic Niche

The neurogenic niche is composed of cell types, molecules and structures that allow for the proliferation and neurogenesis in certain brain areas while preserving others from these processes. Therefore, the SVZ should not be seen as a layer composed of isolated or independent cells, but as a whole. The involvement of blood vessels and the CSF also plays a key role. Many molecules have been identified in the SVZ, some of which are directly related to neurogenesis. This is the case for *noggin*, a protein produced by ependymal cells, which is released to the extracellular matrix and antagonizes another molecule, *BMP4* (bone morphogenetic protein-4), expressed by SVZ astrocytes, and which favors the formation of new neurons while diminishing glia generation (Lim et al. 2000).

This neurogenic niche forms an organized pattern and is widely described in mice. It undergoes variable modifications in different mammals, but the key elements

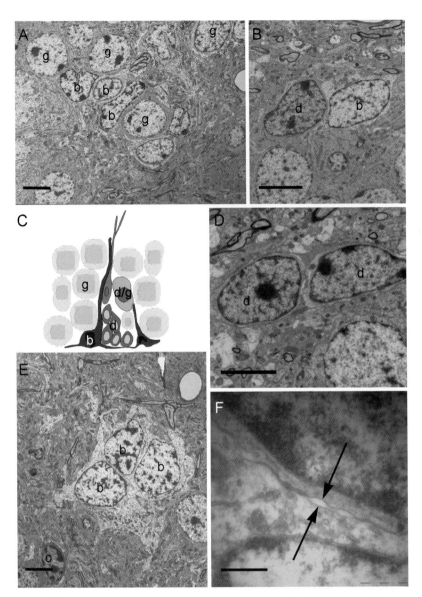

Fig. 21a–f Hippocampal neurogenesis. **a** An active neurogenic niche in the hippocampal dentate gyrus with three B cells (*b*) surrounded by granular neurons (*g*), *scale bar* 5 µm. **b** D cells are the progeny of B cells and migrate very short distances, usually along B cells like the scaffold, *scale bar* 5 µm. **c** Diagram of the hippocampal neurogenic niche with radial stem cells (*blue*) and migratory D cells (*green*) that migrate short distances (*brown*) and differentiate into granular neurons (*pink*). A premature granular cell is shown in *orange*. **d** Type D cells (*d*) are electron-dense with prominent nucleoli and scarce cytosol, *scale bar* 5 µm. **e** An inactive neurogenic niche only consisting of B cells, *scale bar* 5 µm. **f** Intercellular spaces between D cells and D–B cells (*arrows*) are also present, but are narrower than those in Type A cells, probably because migration is not as extensive as in the SVZ, *scale bar* 500 nm (*o*, oligodendrocyte)

required for neurogenesis to occur persist. Therefore, it is highly similar in both other rodent groups and other macrosmatic animals, although it progressively changes in upper mammals such as humans (microsmatic). The niche concept is not specific to the SVZ. There is also a niche in the SGZ, formed by stem cells/progenitor cells, blood vessels, and mitosis.

Blood vessels have also become very important elements, with a large percentage of mitosis associated directly with them, and it has been suggested that they probably play a neurogenic role (Palmer et al. 2000).

3.2
Identification of the Adult Neural Stem Cell in the SVZ

Cells that give rise to neurons during the embryonic development differ from cells that give rise to glia. According to this theory, neuronal stem cells eventually run out and become ependymal cells, while radial glia differentiate into glial stem cells, giving rise to only oligodendrocytes and astrocytes. In the late prenatal and early postnatal stages, these glial stem cells migrate throughout the brain to constitute a glial network and complete myelination. There are two types of mitotically active cells, neuronal precursors and glial precursors, and both coexist in the VZ during the neurogenic peak in the embryo to originate a heterogeneous population in the proliferative region. This hypothesis was confirmed by a study (Levitt et al. 1981) in which both cell types were distinguished according to the expression of a specific marker for glia, GFAP. Mitotically active GFAP$^+$ cells and mitotically active GFAP$^-$ cells would indicate the existence of glial precursors and neuronal precursors, respectively. Nevertheless, these analyses were based on morphological and immunocytologic data and only characterized the proliferative region; they did not test which cell types could be derived in vivo from each precursor. Following the fate of these cells in vivo would lead (and in fact led) to an assessment of whether just one type of progenitor cell could indistinctly give rise to neurons or glia.

The adult NSCs were identified in a similar way in both of the neurogenic regions described in mice. With an elegant experimental design, the group of Alvarez-Buylla (Doetsch et al. 1999) demonstrated that SVZ astrocytes (Type B cells) were the aNSCs of the lateral ventricles. The osmotic infusion of an antimitotic drug (cytosine-beta-D-arabinofuranoside; Ara-C) for six days resulted in the death of all the active proliferating populations in the SVZ. The depletion of C cells was observed in the SVZ in all cases, and only astrocytes and ependymal cells remained. Following this, the SVZ could regenerate completely (Fig. 24). When specific viral tracers or 3HT were injected after Ara-C treatment, astrocytes were labeled and sequentially transformed into C cells, and afterwards into migrating neuroblasts. This lineage progression could be an artifact resulting from the antimitotic treatment. The use of transgenic mice that carried a receptor for an avian virus in their astrocytes solved this problem. An avian virus incorporating the alkaline phosphatase gene was injected and the same sequence of events was observed, thus confirming the previous results.

However, the stem cell in the SVZ generates not only neurons in vivo but also oligodendrocytes (demonstrated primarily in vitro). Type B cells will give rise to not only Type C cells but also the oligodendrocyte precursors (OPCs) that express markers for oligodendroglial differentiation, such as Olig2. A small subpopulation of Type C cells also expresses Olig2 and will, therefore, also originate OPCs. OPCs are Tuj1⁻, PDGF⁺, and PSA–NCAM⁺, in contrast to neuroblasts, which are Tuj1⁺, PDGF⁻, and PSA–NCAM⁺. They migrate toward the CC, where they differentiate terminally into mature oligodendrocytes that express NG2 and synthesize myelin. Demyelinating injury produces a compensatory increase in Type B cells that confers a potential role in myelin repair, a scenario with far-reaching repercussions for therapies for demyelinating diseases (Menn et al. 2006).

Ependymal cells are thought to play an essential role in the movement of the CSF. Some authors have even considered these cells to be capable of proliferating and differentiating into neurons (Frisen et al. 1998). This hypothesis has not been accepted by most of the scientific community because other groups have not yet demonstrated it. An experimental design was performed to relate radial cells observed during postnatal development to neurons and glia. The experiment involved using transgenic animals with a construct under a general promoter (beta-actin) and a *cre-lox* system (see Sect. 2.5 for further details). These animals were injected with a virus which expressed *cre* recombinase in the distal expansions of the radial cells which, on postnatal day 0, would eliminate the *lox–lox* system and allow the appearance of a visible marker. Radial cells and some neurons in the lateral walls of the lateral ventricles were labeled. It was then demonstrated that, during neonatal stages and at short injection times, radial glia expressed different markers depending on the injected virus. The most interesting result obtained is that, as time progressed, not only were labeled neurons found in the OB, but different cell populations were also found in the SVZ. Therefore, it can be stated that radial cells in P0—considered to be glial cells—are the source of neurons, oligodendrocytes, astrocytes, and ependymal cells (Merkle et al. 2004). This work demonstrates that ependymal cells are derived from radial glia, while ependymal cells cannot originate other cell types. In addition, the postmitotic nature of ependymal cells strongly suggests that these cells do not function as adult NSCs in adults. A recent paper by Coskun et al. (2008) discovered that the population of multiciliated ependymal cells was positive for the stem cell marker CD133. CD133⁺ cells have been considered a stage previous to the adult NSCs—as we define them now—because their multipotentiality and oncogenecity are higher. Thus, they propose that the CD133⁺ ependymal cells act as bona fide NSCs in the adult mammalian brain. This is a controversial finding that will require more research evidence before it is accepted.

As previously described, in the adult mammalian DG, new neurons are born in the SGZ and migrate a short distance to differentiate into granule cells (Markakis and Gage 1999; Stanfield and Trice 1988). Using an approach similar to that used to identify the NSCs in the SVZ, the SGZ astrocytes of the DG were also identified as the primary precursors in the formation of new neurons in the adult hippocampus (Seri et al. 2001). First, many proliferating SGZ astrocytes were rapidly converted into a cell type that is GFAP⁻ and which possesses characteristics of D cells. Second, antimitotic treatment with APB (Ara-C plus procarbazol) resulted in the elimination of D cells from the SGZ, but neurogenesis returned and completely regenerated the SGZ after 2–15 days. New neurons were born at a time when 3HT-labeled astrocytes were observed. To support these results, this study showed that SGZ astrocytes that were specifically labeled with the described avian retrovirus gave rise to granule neurons. The retention of 3HT labeling by astrocytes for longer periods of time (Cameron et al. 1993) may be due to symmetric division, an intrinsic characteristic of stem cells. Experiments with viruses and BrdU suggest that radial astrocytes divide to generate D1 cells, which in turn divide once to form postmitotic D2 cells. D2 cells mature through a D3 stage to form new granule neurons, providing a model for the organization of the germinal zone in the adult hippocampus (Seri et al. 2004). The D cells in the SGZ are small and do not seem to divide as frequently. This suggests that the amplification of neuronal production by transient precursors in the SGZ is probably limited. These data do not preclude the possibility that astrocytes could give rise to granule neurons directly without passing through the D cell intermediate. To be able to answer this question appropriately, specific markers to label D cells and methods to directly visualize the conversion from astrocytes to neurons are required. This interpretation was consistent with a previous report suggesting that small, electron-dense cells, similar to D cells, may serve as neuronal precursors (Kaplan and Bell 1984). This conclusion was obtained from independent experimental assays.

3.3
Other Proliferating and Neurogenic Centers in the Adult Brain

The description of new areas with neurogenic potential in nonhuman mammals can be extrapolated—albeit cautiously—to humans, which makes the discovery of new proliferative and neurogenic regions extremely interesting.

In the previously described regions, neurogenesis is well demonstrated, and there is no doubt about its existence. However, studies are emerging on new regions in which neurogenesis may exist, although they are not yet completely accepted. It has been speculated that neurogenesis is present in the neocortex (Dayer et al. 2005; Gould et al. 1999b), in Ammon's horn of hippocampus (Rietze et al. 2000), the caudate nucleus (Luzzati et al. 2006), the amygdala (Bernier et al. 2002), and substantia nigra (Zhao et al. 2003), among others. Other works, however, do not support these results (Koketsu et al. 2003; Kornack and Rakic 2001). Neurogenesis in the cortex is still being debated. Two studies demonstrate the formation of new neurons in the adult cerebral cortex in both mice (Luskin 1993) and primates (Gould et al. 1999b). Nevertheless, some scientists doubt the existence of neurogenesis in the cerebral cortex, as previously mentioned (Kornack and Rakic 2001). In both cases, the presence of new neurons integrating efficiently into the cortex was described, although it was hypothesized that these neurons originated far from their final location or from the precursors that divided and migrated from other areas, probably the SVZ. Thus, it cannot be concluded that these are new neurogenic areas without some controversy. A recent report shows that pericytes, microvascular cells associated with capillaries within the adult CNS, are nestin/NG2$^+$ and behave like stem cells in the presence of bFGF. They can form neurosphere-like structures, self-renew, and are able to differentiate into cells of glial and neural lineages, showing mostly a dual but also a triple lineage potential (Dore-Duffy et al. 2006). Other authors consider neurogenesis to be a result of brain injury (Magavi et al. 2000). It is important to state that the results obtained by different authors have been obtained from different animals, with different treatments, and by means of different experimental designs and/or conditions.

In addition, throughout the ventricular system, it is possible to find cells that proliferate under natural conditions, even when it has not been shown that they can give rise to neurons in vivo (Beattie et al. 1997; Horner et al. 2000; Weiss et al. 1996; Xu et al. 2005). This is the case for the SCR, an expansion of the SVZ that is situated between the cortex and the hippocampus. Another region is the central canal of the spinal cord.

3.3.1
The Subcallosal Region

The SCR derives embryologically from the development of the VZ, which loses its connection at older stages of development and in the postnatal brain. This area consists of clusters of cells that resemble the cellular composition and organization of the SVZ. There are Type A, B, C and E cells, which have identical ultrastructural features (Fig. 22). However, viral injection, in vitro culture, and grafting studies point to a major role in oligodendrocyte (OD) production (Seri et al. 2006). This is not an entirely unexpected finding, considering that the SVZ can also originate oligodendrocytes that migrate toward the corpus callosum (CC). In this particular case, however, it could represent a much greater percentage of the total production of oligodendrocytes. Despite this, neurospheres were obtained from SCR explants in vitro, and their tripotentiality was demonstrated after differentiation.

The chains of migratory cells do not constitute a mesh-like network as they do in the SVZ, but they accumulate in isolated clusters with no contact between them. EM studies of this region reveal that, unlike the RMS, the glia of glial tubes are discontinuously organized, partially ensheathe migratory cells, and allow these cells to make contact with amyelinic axons.

BrdU or 3HT injection at long survival times has not demonstrated neurogenesis in either the CC or hippocampus from resident cells in the SCR under natural conditions. However, it should be noted that such changes after injury remain unknown.

3.3.2
The Central Canal of the Spinal Cord

In recent times, therapy for spinal cord injury has been one of the main interests of the scientific community. Therapeutical approaches vary from the grafting

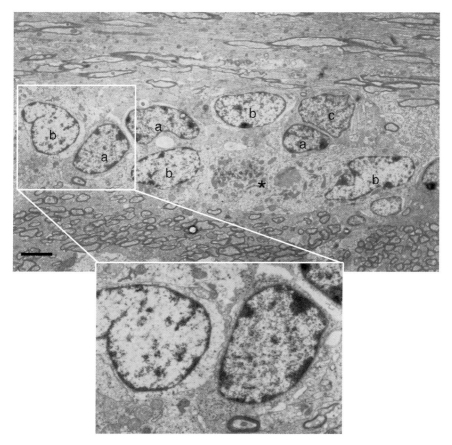

Fig. 22 Subcallosal region. The way the SCR is arranged is comparable to the SVZ with Type B and A cells (*detail*), but it is more densely packed and myelinated. Occasionally, as noted in the picture, there is a virtual LV within the SCR with associated ependymal cells (*asterisk*), *scale bar* 5 μm

of neurons, neuronal progenitors (McDonald 2004) or ensheathing glia from the olfactory bulb, which help to regenerate axons (Nieto-Sampedro 2003; Ramon-Cueto et al. 2000), to the use of biodegradable grafts that facilitate survival and axon regeneration (Lago et al. 2005; Novikov et al. 2002). However, these important advances have not achieved complete functional regeneration. Some authors believe that endogenous stem cells could be potentiated to repair spinal injury if they were to be combined with grafting therapies.

Similar to the lateral ventricles, the central canal of the spinal cord originates from the cavity of the neural tube during embryonic development, so that it derives from the neuroepithelial cells that give rise to mature neurons and glia. Proliferation and even mitotic events have been described in a number of studies (Frisen et al. 1998; Johansson et al. 1999). Mitotic figures that are separated from the canal and immersed in the nervous parenchyma have also been observed. A small fraction of the cells continue to divide after birth (Adrian and Walker 1962), giving rise to glial cells but not neurons (Horner et al. 2000; Johansson et al. 1999). This proliferative capability increases with the exogenous administration of growth factors or after spinal cord injury (Beattie et al. 1997). Some factors are known, such as *Pax6* (Yamamoto et al. 2001), but the cell type that proliferates and the fine cellular organization of the canal both remain unknown. Likewise, the cells of the central canal of the spinal cord in adult rats preserve the capacity to proliferate and differentiate in vitro into neurons, oligodendrocytes, and astrocytes (Kalyani et al. 1998; Shihabuddin et al. 1997; Weiss et al. 1996). Experiments were performed which demonstrated the formation of new neurons from spinal cord stem cells transplanted into the hippocampus (Shihabuddin et al. 2000), indicating that these cells preserve, at least in part, the neurogenic program. In vitro experiments, however, attribute this induction of neurogenesis to signals from astrocytes that originally existed in the adult neurogenic areas (Song et al. 2002). These results have to be interpreted carefully, since the neurosphere assay does not completely reproduce the in vivo pattern and conclusions might be misleading.

Some anamniotes preserve the capability to regenerate the spinal cord after injury, since the ependymal and periependymal layers proliferate and give rise to neurons and glia. In the eel, complete reconstruction is possible even after complete sectioning (Dervan and Roberts 2003a, b). In addition, precursor-like cells and immature neurons have been reported to form the spinal cord in turtles (Russo et al. 2004). After studying adult neural stem cells in the ventricles from birds and reptiles, similarities to mammals have been described (Garcia-Verdugo et al. 2002). Thus, it is crucial to research the regenerative potential of the spinal cord in mammals and the growth factors that inhibit it compared to other animal groups. It is unclear whether it is possible to reestablish the regenerative potential of the stem cell of the central canal of the spinal cord that is apparently lost after birth.

In addition, very little is known about stem cell markers. Some proposed candidates are GFAP or S-100 (Horner and Gage 2000; Horner et al. 2000). Previous studies from our group (unpublished results) describe the existence of multiciliated ependymal cells that do not proliferate, and uniciliated cells with unique radial expansions that proliferate. These uniciliated cells may correspond to nestin$^+$ astrocytes, which could be stem cells. Other cells in the central canal are 3CB2$^+$ (Shibuya et al. 2003).

At an ultrastructural level, very few studies have been published that study the spinal cord. The cells have been described as being ependymal cells with a radial expansion that reaches the blood vessels nearby, which some authors have named tanycytes. These cells have large union complexes, intermediate filaments, microvellosities, and abundant mitochondria. Their nuclei are round with clumped chromatin (Fig. 23). Typical astrocytes can be observed close to the central canal. These astrocytes have abundant intermediate filaments and highly irregular contours. Occasionally, microglial cells and blood vessels are seen.

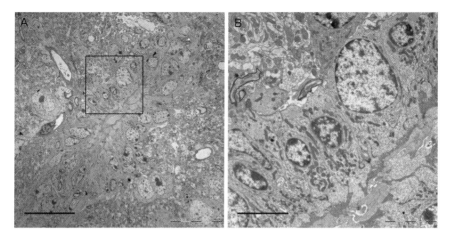

Fig. 23a–b Spinal cord. **a** The central canal of the spinal cord is composed of cubic cells that isolate the CSF from the nervous parenchyma, with uniciliated astrocyte-like cells interspersed among them, *scale bar* 20 μm. **b** A detail of ependymal cells showing large nuclei with clumped chromatin, *scale bar* 5 μm

3.4 Distinct Features of Different Species: Comparative Study of Mice and Humans

Cell proliferation in the SVZ has been demonstrated in many vertebrate species including mice, rats, rabbits, voles, dogs, cows, monkeys, and humans (Blakemore 1969; Blakemore and Jolly 1972; Kornack and Rakic 2001; Lewis 1968; McDermott and Lantos 1990; Rodriguez-Perez et al. 2003; Sanai et al. 2004), while adult hippocampal neurogenesis has also been demonstrated in birds (Barnea and Nottebohm 1994), reptiles (Lopez-Garcia et al. 1992), rodents (Altman and Das 1965), and primates, including humans (Eriksson et al. 1998; Gould et al. 1997).

In this section, we attempt to describe the morphological features of the SVZ in certain species of mammals in comparison with those previously shown in rodents, with a view to highlighting the similarities and differences between them (see the comparative diagram in Fig. 27).

3.4.1 Bovine Lateral Ventricles

Three regions have been described in the bovine lateral ventricles. Region 1 characteristically presents white matter and consists of ciliated cubical ependyma, a few subependymal cells, and a narrow subjacent glial layer. This region probably corresponds to the dorsolateral corner of the SVZ. Region 2 lines the striatum and possesses ependymal cells with basal processes that extend through a layer of dense

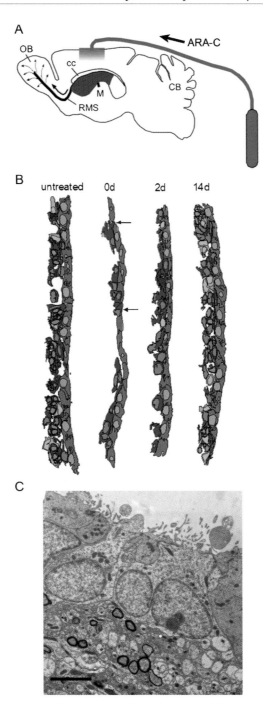

subependymal cells and a wide glial network underneath. Region 3 corresponds to the rostral horn and presents the large expansion of the subependyma and associated glia (Fig. 25).

Immunohistochemically, the bovine SVZ shares common features with rodents. Ependymal cells can be detected with antibodies against S-100b, vimentin, and GFAP (with the exception of ependymal GFAP⁻ cells in Region 1). In contrast to what happens in mice, ependymal cells are positive for nestin and BLBP (brain lipid-binding protein). The small cells present in the subependyma and immersed in the glial network stain for Tuj1, while the glial network underneath shows immunoreactivity for vimentin, but not for GFAP. Proliferation in the lateral bovine ventricle was assessed by PCNA immunostaining. PCNA$^+$ nuclei were abundant in the subependymal and glial layers of the Type 2 and 3 walls. Migration was demonstrated by DiI in vitro tracing studies that revealed small bipolar cells in the glial layer at a distance from the site of the label deposit. Young neuroblasts may migrate in a rostroventral direction through the glial network, but PSA–NCAM demonstration could not be performed because there was no accurate PSA–NCAM antibody that worked in the bovine brain.

These results suggest, but do not prove, that neurogenesis takes place in adult bovine subependyma, mostly in the walls of the striatum and the rostral horn. In addition, cultures from subependyma-derived cells (in the absence of growth factors) generated neurons, suggesting that at least precursor cells were present in the subependyma of explants (Perez-Martin et al. 2000, 2003).

3.4.2
Rabbit Lateral Ventricles

While the ultrastructural features of the rabbit subependyma are almost indistinguishable from the rodent SVZ, the rabbit brain shows certain peculiarities. Because of the persistence of an open olfactory ventricle throughout life (Leonhardt 1972), the SVZ rostral extension (or RMS) of rabbits maintains direct contact with the ependymal cell layer. Remarkably, the neuroblasts in rabbits lack the intercellular clefts that characterize the rodent RMS, and also migrate independently from the astrocytes by sporadically contacting with them, whereas they establish frequent

Fig. 24a–c Ara-C and the discovery of the identity of neural stem cells. Ara-C is an antimitotic drug capable of killing fast-proliferating precursors. After a 6-day infusion of Ara-C into the lateral ventricles (**a**), the SVZ is devoid of Type C and A cells (**b**), the two active proliferating populations (*red* and *green* in the diagram). The only remaining cells are astrocyte-like cells (B cells in *blue*) that proliferate very slowly, and ependymal cells (*gray*) that do not divide. **c** After a few days (14 days in the diagram), the SVZ is completely regenerated. For this to happen, Type B cells have to activate and rapidly generate Type C and A cells. The process of B-cell activation implies contact with the ventricle. After Ara-C infusion, the number of Type B cells touching the ventricle increased significantly (also depicted as *arrows* in **b**), *scale bar* 5 µm

synaptic contacts (Luzzati et al. 2003). Migratory chains composed of PSA–NCAM⁺ Type A cells are distributed in different patterns depending on the brain region. They are arranged in a similar way to the chains in mice close to the ventricular SVZ, forming clusters of cells immersed in an astrocytic meshwork that are less dense than those in rodents. There is a region that authors call the abventricular SVZ. It is separated from the ventricle by a band of astrocytic expansions, which is poor in cell bodies, and which resembles the hypocellular gap layer found in the human SVZ (see Sect. 3.4.4). It shows larger migratory chains and even a less dense astrocytic compartment. Finally, and externally to the abventricular SVZ, migratory chains penetrate the brain parenchyma ("parenchymal chains"), probably as a result of transient migration toward the striatum and the subcortical regions. GFAP immunoreactivity in the astrocytes from different regions varies. Indeed, the further they are from the ventricle, the stronger they are (Ponti et al. 2006) (Fig. 25).

3.4.3
Primate Lateral Ventricles

To date, a detailed description of the primate SVZ has not been provided. Proliferation has been demonstrated using proliferative markers. Migratory chains of PSA–NCAM and Tuj1⁺ cells have also been described. In Old World primates (*Macaca mulatta*), some studies have demonstrated that labeled cells were present in the SVZ, olfactory tract (OT) and OB 2 h after BrdU injection (Pencea et al. 2001a). Most of the BrdU⁺ cells in the SVZ were located ventrally and laterally in the region corresponding to the caudatum. In the olfactory tract, labeled cells elongated to follow the axis of the tract. The fact that proliferating cells were found in the olfactory bulb

Fig. 25a–d Nonhuman mammalian SVZ. A semithin (**a**) and ultrathin section of a detail (*squared box*) (**b**) from the bovine LV. The bovine ventricular zone is a layer of cubical cells constituting the multiciliated ependyma. Under the ependymal layer, multiple undifferentiated cells were found with the morphology of astrocytes and neuroblastic cells. They both shared ultrastructural commonalities that made them very difficult to distinguish. This is explained by the undifferentiated states of both populations. Ependymal radial glia sent expansions toward the neuropile (*arrows*), *scale bar* 5 μm. **c** Diagram depicting the features of the rabbit SVZ. **c1, c2** The olfactory ventricle remained open throughout life (*ov*), so that the RMS of rabbits is in constant, direct contact with the ependymal cell layer. Neuroblasts in rabbit can migrate to form chains in the boundaries of the SVZ, but beyond the SVZ, they do so independently of the astrocytes, and sporadically come into contact with them. There is an abventricular SVZ (*A*) that shows larger migratory chains and an even less dense astrocytic compartment. Migratory chains penetrate the brain parenchyma (*P*) externally to the abventricular SVZ. **c3, c4** Close to the ventricular SVZ, cells are arranged in a similar fashion to the chains in mice, and form clusters of cells immersed in an astrocytic meshwork. **d** The monkey SVZ is composed of four layers: ependymal layer, GAP layer, astrocyte ribbon, and deep parenchyma, *scale bar* 5 μm (*P*, parenchymal chains; *A*, abventricular SVZ; *V*, ventricular zone; *lv*, lateral ventricle; *ov*, olfactory ventricle; *Bv*, blood vessel; *n*, neuron; *m*, microglia). Figure 25c has been modified from Ponti et al., 2006

Primate Lateral Ventricles

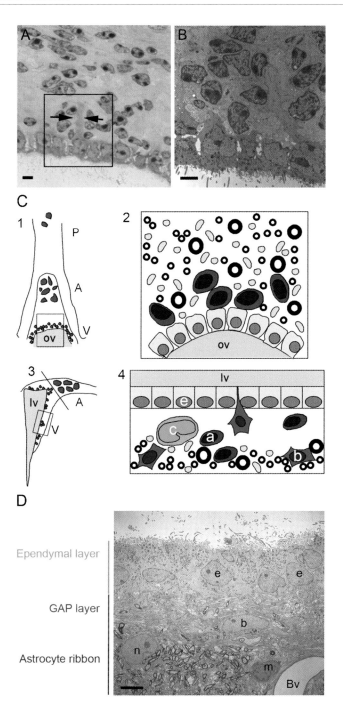

confirms that resident cells retain proliferative potential in monkeys. These results have been contrasted with other markers, such as PCNA. In our studies in the monkey SVZ, ependymal cells were never positive for Ki-67 (Pencea et al. 2001a). Little is known about the ultrastructural features of the SVZ in macaque. As previously described for humans (Quinones-Hinojosa et al. 2006; Sanai et al. 2004), we observed that the SVZ in monkeys is composed of three layers: (a) an ependymal layer, (b) a GAP layer (a hypocellular layer described in the human SVZ), (c) an astrocyte ribbon, and finally the deep parenchyma (Fig. 25). Our results show four distinct regions in the rostrocaudal axis (frontal, body, temporal, and occipital), based on the presence of different cell types, the width of the GAP layer, the amount of myelin, and other distinct features. In the frontal region, the GAP layer is wide, myelin is scarce, and clusters of cells with dark nuclei are randomly distributed within the parenchyma. These groups of cells are closely related to blood vessels and occasional mitotic figures can be observed. Conversely, in the body of the ventricle, the GAP layer is variable in width but not as wide as in the frontal region. There are also dark-nuclei cells but they are present in smaller amounts, and they did not form chains. The occipital region presents higher amounts of myelin. There are fewer cells, the GAP layer is narrow, and we do not find dark-nuclei cells. Finally, the temporal region can be subdivided into two well-distinguished regions: the dorsal side of the temporal horn (or hippocampal SVZ), and the ventral side, which is closely related to the hippocampus. No dark-nuclei cells were observed either in temporal regions (Fig. 26).

The three types of cells described in mice are also present in monkeys: ependymal cells, Type B cells (GFAP$^+$), Type A cells (elongated, Tuj1 and PSA–NCAM$^+$). It is not clear, however, whether Type C cells (BrdU$^+$, negative for A- or B-cell markers) exist in primates. Confocal images have shown that these cells are arranged similarly in mice and form chains of Type A cells surrounded by astrocytic expansions. These chains run along the monkey RMS (vertically and horizontally) within the OT, and converge in the OB. The generation rate, migration and/or differentiation in monkeys are slower than in mice, and it takes up to 75 days to visualize BrdU$^+$ neurons in the OB (Kornack and Rakic 2001). Based on our work, we consider the decrease in the number of migratory cells along the pathway to be interesting, because it may imply that significant cell death is occurring (Gil-Perotin et al., unpublished results).

3.4.4
Human Lateral Ventricles

The existence of neurogenesis in the hippocampus has been demonstrated, although it remains unclear whether neurogenesis occurs in the human OB, since the availability of material is restricted or too old. Hippocampal neurogenesis was shown in postmortem tissue obtained from cancer patients who had been treated with BrdU. Double labeling for both BrdU and a neuronal marker (NeuN, calbindin, or NSE) was found in the dentate gyrus of adult humans (Eriksson et al. 1998). Occasionally,

BrdU⁺ cells in the SVZ did not express neuronal markers. It should be pointed out that OBs were not included in these studies.

So far, it has been accepted that the walls of the lateral ventricles host cells with astrocytic features that proliferate and differentiate into neurons, astrocytes, and oligodendrocytes under certain culture conditions. These cells have been identified as being the adult NSCs in humans (Sanai et al. 2004), and they resemble the ones studied in the rodent brain. The cytoarchitecture of the SVZ, however, differs a great deal from that described in mice. The SVZ can be divided into three distinct layers:

1. A unique layer composed of cubical ependymal cells in contact with the ventricular lumen. These cells send radial expansions into the glial network (neuropile) and the second layer.
2. A hypocellular layer, or GAP layer, with fewer cells, basically formed from a large number of ependymal and astrocytic expansions, the latter being rich in intermediate filaments. Occasional neurons and displaced ependymal cells can be observed. These ependymal formations found far from the lumen of the ventricle are exclusive to this layer, and are composed of between 5 and 20 cells. These cells are ultrastructurally identical to ependymal cells, with long union complexes among them. They are arranged to form sphere-like structures with cilia and microvilli interacting toward the center.
3. Layer 3, or the astrocyte ribbon, is where astrocytes bodies that send expansions to the GAP layer reside. This layer comprises large astrocytes, myelinic axons, oligodendrocytes, and a progressive increase in synaptic contacts. These astrocytes have been identified as adult NSCs. Unlike rodents, these stem cell-like astrocytes are separated from the ependyma by the GAP layer. An interesting fact is that these cells sporadically send expansions that contact the ventricle, and which resemble those observed in the SVZ of mice. Underneath the astrocyte ribbon there is a transitional zone in the brain parenchyma that corresponds to the presence of neuronal bodies.

Three different types of astrocytes were found, based on ultrastructural morphology. In the GAP layer, occasional astrocytes (B1) were small and very rich in intermediate filaments. In the astrocyte ribbon, the second type of astrocytes (B2) was characterized by being large and rich in organelles and intermediate filaments in their cytosol. Type B3 is found in the transitional zone between layer 3 and the brain parenchyma. Within the parenchyma they correspond to large astrocytes that are poor in cytosolic organelles or intermediate filaments (Fig. 26). Interestingly, the astrocytes of the ribbon sent long prolongations to make contact with the lumen of the ventricle. It remains unknown whether this expansion emits a primary cilium, as in mouse.

Proliferating cells (BrdU⁺, mitosis, Ki-67⁺) were always found far from the ependymal layer, and these cells were always GFAP⁺ cells of the astrocyte ribbon (Quinones-Hinojosa et al. 2006; Sanai et al. 2004). Cells derived from the human SVZ (expressing GFP under the GFAP promoter) were processed and enriched with FACS analysis to separate GFAP⁺ (fluorescent) from GFAP⁻ cells. The GFAP⁺ cells

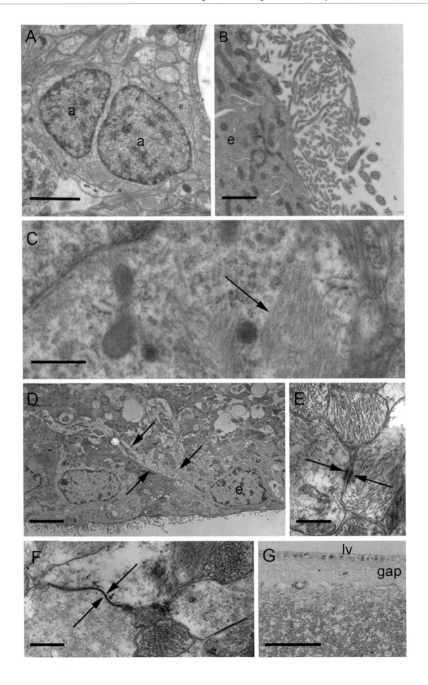

from the astrocyte ribbon, but not from the cortex, were the only ones capable of forming neurospheres in vitro. The more enriched the medium was in GFAP⁺ cells, the more neurospheres were obtained, thus demonstrating that astrocytes were capable of proliferating in response to growth factors, and resembling the neurosphere assay in mice. Under differentiation conditions, these neurospheres gave rise to oligodendrocytes, astrocytes, and neurons, indicating that these cells were the neural stem cells in the adult human SVZ. It is still a challenge to establish a map describing the localization and concentration of stem cells in the human SVZ.

It was only in the anterior levels that elongated Tuj1⁺ cells with a migratory morphology were described, and the adult human SVZ appears to be devoid of chain migration. This could be the result of either a lack of quality in the human material or the age of the individuals from whom the tissue came, but could also be due to the degree of neurogenesis in the OB being low in humans compared to rodents. It would be easier to find migratory cells at younger ages. The lack of tissue samples from humans and their conservation are responsible for what has been finally observed in EM, and can lead to mistaken interpretations. Migratory cells to the OB may exist, but specific molecular markers cannot bind and EM morphologic criteria cannot be assessed in damaged tissue. A recent paper states that chain migration takes place toward the human OB and that the lumen of the ventricle expands until the OB (Curtis et al. 2007). This paper has been criticized by Alvarez-Buylla's group, who defend the notion of the absence of ventricle lumen in the OB and migration toward the human OB, which, even if it did exist, would not resemble that in rodents (Sanai et al. 2007).

Comparative studies in different primate groups would probably advance knowledge of the human SVZ.**Identification of the Adult Neural Stem Cell in the SVZOther Proliferating and Neurogenic Centers in the Adult BrainDistinct Features of Different Species: Comparative Study of Mice and Humans**

Fig. 26a–g Monkey and human SVZ. The monkey (a–c) and human (d–g) SVZ share many commonalities. They are commonly arranged into four layers, for instance, in both species. The main difference found by our group was that we did not find the RMS in the human brain. **a** The migratory cells close to a blood vessel in *Macaca fascicularis*, *scale bar* 2 μm. **b** Ependymal cells with abundant microvellosities and cilia, *scale bar* 1 μm. **c** Intermediate filaments are abundant in the cytosol of ependymal cells, *scale bar* 500 nm. **d** In humans, ependymal cells display radiality and send expansions toward the brain parenchyma, which are highly similar to those described in the bovine and monkey brain (*arrows*), *scale bar* 5 μm. Note the lower quality of tissue samples derived from the human brain because of a delay in fixation. **e** Specialized cell-to-cell GAP junctions in the GAP layer between astrocytes and ependymal cells (*arrows*), *scale bar* 200 nm. **f** Zonula occludens between astrocytes and ependymal cells (*arrows*), *scale bar* 200 nm. **g** A semithin picture showing the width of the GAP layer (*gap*)—very large compared to other species, *scale bar* 50 μm

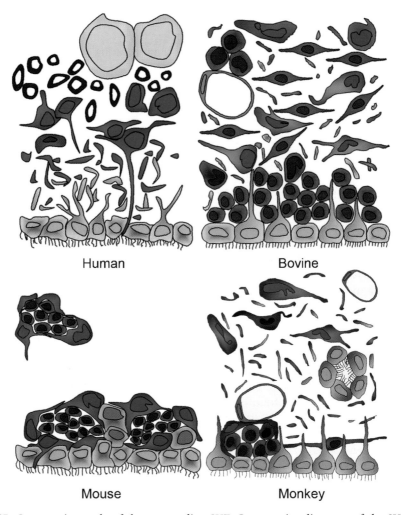

Fig. 27 Comparative study of the mammalian SVZ. Comparative diagrams of the SVZ in four distinct species of mammals. See text for further details (*orange*, ependymal cells; *blue*, astrocytes; *green*, Type C cells; *red*, migratory neuroblasts; *violet*, mature neurons)

4
Oncogenesis vs. Neurogenesis

The histological classification (WHO) of brain tumors sorts them according to their resemblance to a specific cell lineage, yielding glial tumors (gliomas or oligodendrogliomas), neuronal neoplasms (central neurocytoma, ganglioglioma, neuroblastoma), or tumors derived from ependymal cells (ependymoma). Among these, the most aggressive glioma, the glioblastoma multiforme (grade IV), has the distinctive features of palisading or geographic necrosis and conspicuous microvascular hyperplasia in addition to marked cellular pleomorphism. Glioblastoma multiforme offers a poor prognosis, even after complete treatment and resection.

A question that arises from the existence of NSCs and glial progenitors throughout the adult brain is whether these cells could become malignant under certain conditions and thus be the originating cell for brain tumors. This hypothesis has not been scientifically proven or widely accepted. The proliferative and undifferentiated phenotype that characterizes stem cells suggests that the deregulation of their function could result in uncontrolled growth (Fig. 28). There is recent evidence that certain brain tumors contain small numbers of cells with NSCs properties, which include the ability to self-renew, proliferate and differentiate (Singh et al. 2003). The SVZ has been linked to the formation of glioma in animal models (Gil-Perotin et al. 2006; Hopewell 1975; Lantos et al. 1976; Vick et al. 1977). The SVZ is a neurogenic niche; it is a region where cell growth occurs. The cytoskeleton, cell adhesion molecules, and growth factors (Table 4.1) allow cell proliferation, and may be instrumental in glioma development. This is the case for nestin, an intermediate filament that is expressed widely by stem cells at various differentiation stages, but is also found in human glioma, thus providing further support for the link between SVZ stem cells and glioma formation (Dahlstrand et al. 1992; Uchida et al. 2004). Dcx, a cytoskeleton component that is used as a marker for young neuroblasts in the RMS, is expressed in high-grade gliomas in relation to migration and invasiveness (Daou et al. 2005). Cadherins and catenins, cell adhesion molecules expressed in the SVZ, have also been related to brain tumors when altered, and mostly correlate with tumor dissemination (Asano et al. 1997, 2004; Barami et al. 2006; Perego et al. 2002; Utsuki et al. 2002).

There are two types of glioblastoma, which is classified according to its origin, since it can arise as a progression from astrocytomas of a lower grade or it can be newly generated. The former requires a synergism between distinct pathways,

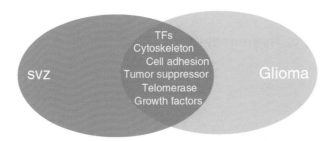

Fig. 28 Gliomagenesis and neurogenesis. The SVZ is a proliferative region that contains adult NSCs. These cells retain the ability to proliferate, as they are uncommitted neural precursors. The environment (neurogenic niche), including blood vessels, supplies growth factors from neighboring or distant cells, cell adhesion molecules and cytoskeletal scaffolding, facilitating all of the processes implied in proliferation, differentiation, and migration. On the other hand, the genetic program of the cell is also permissive in regard to proliferating and tumor suppressor genes, telomerase activity, and transcriptional factors. Some of these elements that are necessary for neurogenesis are altered in tumor cells, and thus beg the question of whether adult NSCs are the originating cells for brain neoplasms. This crossroads where both populations meet has been the focus of numerous recent reports (see text)

Table 4.1 Growth factors expressed in the SVZ and gliomagenesis

Growth factor	Gliomagenesis when	References
EGF	EGFR amplification	Wechsler-Reya et al. (2001)
PDGF	Excessive PDGF activation	Assanah et al. (2006) and Jackson et al. (2006)
IGF-1	Overexpressed in glioma cell lines	Trojan et al. (1993)
VEGF	Proangiogenic factor expressed during glioma growth	Plate et al. (1994)
TGFb	Increased in plasma of glioblastoma multiforme patients	Schneider et al. (2006)

such as the loss of p53 after increased PDGF (platelet-derived growth factor) signaling (Hesselager et al. 2003), prenatal exposure to ENU (*N*-ethyl-*N*-nitrosourea) (Leonard et al. 2001; Oda et al. 1997), or Ras activation (Bajenaru et al. 2001; Reilly et al. 2004). Early inactivation of p53 has been shown to cooperate with the neurofibromatosis-1 (NF1) tumor suppressor gene mutation to induce malignant astrocytoma formation in mice via the constitutive activation of Ras signaling (Zhu et al. 2005). A recent publication from our group demonstrates that the absence of p53 not only increases susceptibility to glial tumors, but it also directly affects the biology of NSCs in the adult SVZ by conferring an advantage in proliferation, increasing self-renewal capability without transforming the cells, and preserving their potential to terminally differentiate (Gil-Perotin et al. 2006). When the loss of p53 is associated with a mutagenic stimulus, this leads to enhanced self-renewal of the SVZ NSC cells, recruitment to the fast-proliferating compartment, and impaired differentiation (Fig. 29a, b).

The alternative, de novo tumorigenesis, is represented by the activation of the EGF receptor (EGFR) in astrocytes. The expression of constitutively active EGFR associated with the loss of other cell cycle genes such as the INK4α/ARF locus (Bachoo et al. 2002) is sufficient for glioblastoma formation.

PTEN (phosphatase and tensin homolog deleted on chromosome 10) has also been acknowledged to be a tumor suppressor that is mutated in gliomas (Li et al. 1997).

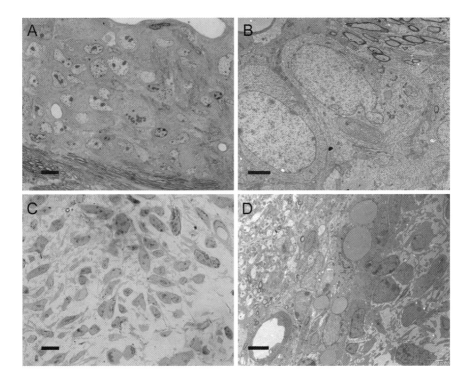

Fig. 29A–D Gliomagenesis and SVZ. **A, B** Gliomagenesis in p53 mutant mice after a mutagenic stimulus. After ENU exposure during pregnancy, adult mice lacking p53 protein expression develop glioblastoma-like tumors close to the lateral ventricles. **a** Semithin section of the tumor that is interposed between the corpus callosum (CC) and the lumen of the ventricle. At this magnification (LM), two cells are in mitosis, and tumor cells display large invaginated and atypical nuclei, *scale bar* 10 μm. **b** Detail at high magnification (EM), showing cells with ultrastructural characteristics that resemble the Type C cell of the SVZ but which are more atypical. Microtubules or intermediate filaments were not present in their cytosol, *scale bar* 1 μm. **C, D** Gliomagenesis de novo with a constitutive PDGF stimulus. Mice carrying an osmotic pump with a PDGF infusion for 15 days developed tumors in the SVZ. These cells seemed to be less differentiated, as seen in the embryo during development, with large intercellular spaces and a fibroblastic appearance. **c** *Scale bar* 10 μm, **d** *scale bar* 5 μm

It acts as a phosphatase that lowers the levels of PIP3 (phosphatidylinositol triphosphate) and enhances the rate of apoptosis. PTEN is expressed in SVZ precursor cells (Li et al. 2002) and regulates neuronal differentiation and migration. It negatively regulates NSC proliferation.

PDGF signaling is particularly interesting, as human gliomas express a subtype of the PDGF receptor, as do glial progenitors and neural stem cells (Jackson et al. 2006; Shoshan et al. 1999). Infusion of PDGF in the lateral ventricles gives rise to the self-limited outgrowth of cells in close relation to the lateral ventricles (Jackson et al. 2006) (Fig. 29c, d). In addition, the retroviral expression of PDGF in glial progenitors throughout the brain is capable of generating neoplasms that highly resemble glioblastoma in the neonatal (Dai et al. 2001; Uhrbom et al. 1998, 2000) and adult (Assanah et al. 2006) brain. Therefore, it is currently being discussed as to whether the autocrine/paracrine effect exerted by the PDGF ligand occurs in the Type B cell, in the bona fide stem cell in the SVZ, in the more restricted glial progenitors, or in the OPCs localized throughout the brain, and more extensively in the subcortical white matter.

In short, key components of the molecular regulation of tissue and cancer stem cell features may be shared, and tumor formation, in at least some aspects, can be viewed as excessive stem cell expansion (Pardal et al. 2005). In fact, it has been demonstrated that a marker for human brain stem cells, CD133, is expressed by the most aggressive and self-renewing fraction of human glioblastoma cells in vitro (Singh et al. 2003; Uchida et al. 2000).

A hot topic in stem cell biology has been the migration of the newly formed cells toward target regions. In mice, this question has been studied thoroughly and, as previously described, neurogenesis is concentrated in both the OB and hippocampal DG. The restricted glial progenitors, which are thought to be derived from Type C cells (Menn et al. 2006), migrate toward the white matter—mostly the CC—and remain as OPCs to be recruited for the remyelination of injured tracts. Not only are SVZ stem cells able to give rise to gliomas after undergoing malignant transformation, but they can also be attracted toward gliomas and contribute to the overall mass effect of the developing glioma (Aboody et al. 2000). They may even exhibit antitumor effects (Arnhold et al. 2003; Glass et al. 2005). This tropism is currently under study (Bao et al. 2006; Tabatabai et al. 2005; Visted et al. 2003; Ziu et al. 2006), but the environmental cues and mechanisms that determine this centripetal migration and the reverse centrifugal migration of tumor cells toward the brain parenchyma remain unclear.

How glioblastomas are generated or how they migrate are questions to be answered in the future. The SVZ lies at the intersection of brain development and gliomagenesis. Increasing our knowledge of stem cell biology and the regulatory mechanisms for proliferation, differentiation and cell death will shed light on the origin of brain tumors, and the identification of the originating cell for gliomas will allow the creation of more relevant animal models. Furthermore, it will help us to devise more effective therapies that will allow us to, if not prevent or completely cure brain cancer, at least diminish its social impact.

5
Adult Neurogenesis Under Pathological Stimulation: Ischemia

5.1
Concept and Epidemiology

There is extensive literature that analyzes the factors involved in regulating proliferation and neurogenesis (Tables 5.1 and 5.2), and it is widely accepted that most of the pathologic brain states influence and regulate both processes, especially in the SVZ. It has been demonstrated that brain injury comprising traumatisms (Carmichael 2003; Rola et al. 2006), ischemia (stroke) (Tonchev et al. 2005), neurodegenerative diseases (Curtis et al. 2003; Jordan et al. 2006; Sonntag et al. 2005), epilepsy (Parent et al. 2002, 2006), and gamma-irradiation (Balentova et al. 2006; Uberti et al. 2001) triggers changes in the properties and structure of the neurogenic areas of the adult brain. We now go on to briefly review the effect of ischemia on neurogenesis, given its social impact.

Brain ischemia occurs when an artery to the brain is blocked. The term *ischemic stroke* refers to a sudden endangering of cerebral functions due to a variety of histopathological alterations involving one or more blood vessels. Blood carries oxygen and nutrients to the brain, and takes away carbon dioxide and cellular waste. If an artery is blocked, brain cells (including neurons) cannot make enough energy and will eventually stop working. If the artery remains blocked for more than a few minutes, brain cells may die.

Acute cerebral vascular, ischemic and/or hemorrhagic disease is the third cause of mortality and the first cause of disability in most Western countries. Ischemic stroke is by far the most common form of stroke, accounting for around 88% of all strokes. Stroke can affect people of all ages, including children. Many people with ischemic strokes are older (60 years old or more), and the risk of stroke increases with age. At each age, stroke is more common in men than women. A high proportion of the people who survive strokes suffer side effects in the form of mental or physical problems, and require permanent help in their daily activities. Although the death rate due to stroke, and maybe its incidence, has dropped in recent decades (Bonita and Beaglehole 1993), it is possible that at least the incidence rate will increase in the near future (Broderick et al. 1989; Jorgensen et al. 1992; Terent 1988). This will happen, among other reasons, because people are now more likely to survive into old age than ever before, and this is leading to an aging population,

Table 5.1 Factors affecting proliferation of adult NSCs and/or progenitors

Factor	Effect	References
Genetic background C57/BL (h)	↑	Kempermann et al. (1997)
Adrenalectomy (h)	↑	Yehuda et al. (1989)
Epilepsy (h)	↑	Parent et al. (1997)
Traumatism (48 h after injury) (h)	↑	Rola et al. (2006)
Low dose X-irradiation (h)	↓	Mizumatsu et al. (2003)
Ischemia (s)	↑	Tonchev et al. (2005)
FGF2 (s)	↑	Wagner et al. (1999)
EGF (s)	↑	Kuhn et al. (1997)
TGFa (s)	↑	Cooper and Isacson (2004)
BDNF (s)	↑ (p75 receptor)	Zigova et al. (1998)
VEGF (h, s)	↑	Jin et al. (2002b)
Ephrins (s)	↑	Conover et al. (2000)
TNFa (s)	↑	Wu et al. (2000)
BMP (s)	↓	Coskun et al. (2001)
CNTF/LIF (s)	↑ Self-renewal	Shimazaki et al. (2001)
Shh (h, s)	↑	Charytoniuk et al. (2002) and Palma et al. (2005)
Notch (s)	↑	Chambers et al. (2001)
Tenascin C (s)	↑ Self-renewal EGFR responsiveness	Lillien and Raphael (2000)
Serotonin (h, s)	↑	Banasr et al. (2004)
Dopamine (s)	↑	Baker et al. (2004) and Coronas et al. (2004)
Estrogen (h)	↑	Gould et al. (2000) and Saravia et al. (2004)
Abeta (s)	↓	Haughey et al. (2002b)
Ethanol (h)	↓	Crews et al. (2003)

in which ictus is more frequent. Therefore, ischemia will continue to be an important sanitary and economic burden on health systems, not to mention the personal, social, family, and working problems caused by it. It should be noted that ictus produces the highest number of admittances to neurological services, and it is the neurological disease with the highest risk of death. Furthermore, ictus consumes 5% of the Scottish National Health Service budget (Forbes 1993). In view of this sanitary scenario, it is obvious that we are faced with a very serious problem, and that it is important to look for solutions in primary prevention given the potential they have to reduce the effect of this disease.

Experimental investigations of focal transitory cerebral ischemia are important, since it is the equivalent to human transitory cerebral ischemia. These results suggest

Table 5.2 Factors affecting neurogenesis of adult NSCs and/or progenitors

Factor	Effect	References
Genetic background 129/SvJ (h)	↑ gliogenesis	Kempermann and Gage (2002); and Kempermann et al. (1997)
Epilepsy (h)	↑ neurogenesis	Parent et al. (1997)
Low dose X-irradiation (h)	↓ neurogenesis	Mizumatsu et al. (2003)
Traumatism (h)	↓ neurogenesis ↑ gliogenesis	Rola et al. (2006)
FGF2 (s)	↑ neurogenesis	Wagner et al. (1999)
EGF (s)	↓ neurogenesis ↑ gliogenesis	Doetsch et al. (2002) Kuhn et al. (1997)
BDNF (s)	↑ neurogenesis ↑ survival	Pencea et al. (2001a, b) Kirschenbaum and Goldman (1995)
EPO (s)	↑ neurogenesis	Shingo et al. (2001)
HB-EGF (s)	↑ neurogenesis	Jin et al. (2002a)
BMP (s)	↑ gliogenesis	Gross et al. (1996)
Noggin (s)	↑ neurogenesis	Lim et al. (2000)
CNTF/LIF (s)	↑ neurogenesis	Emsley and Hagg (2003)
Serotonin (h, s)	↑ neurogenesis	Banasr et al. (2004)
Dopamine	↑ neurogenesis	Van Kampen and Robertson (2005)
Abeta (s)	↓ neurogenesis	Haughey et al. (2002a)

that a 5-min transitory focal ischemia with a 7-day recovery is enough to cause modifications in the morphology and neural density of the hippocampus or striatum. When the cerebral tissue is damaged by acute ischemia, the interruption of the blood flow to the brain produces a fast alteration in the metabolism, which leads to two forms of damage. The first will be immediate and irreversible damage, called necrosis zone or core. It takes place in the area surrounding the occluded vessel. The second, which takes place later, is a time-dependent zone; in a matter of minutes, hours, or even days, the parenchyma will progressively be destroyed as a consequence of a complex succession of biochemical alterations (Fig. 30). This is called the penumbra zone. The specific abilities that will be lost or affected by stroke depend on the extent of the brain damage and—most importantly—where the stroke actually occurred in the brain.

5.2
Effects of Ischemia on the Brain and the SVZ

When the blood flow is interrupted in the focal ischemia model, the brain tissue needs to initiate endogenous protective processes to avoid cell death. The main processes are the activation of macroglial cells (astrocytes), recruiting of microglial

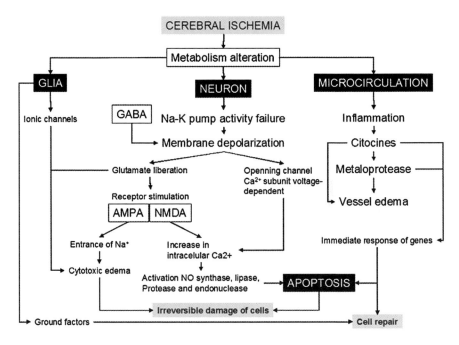

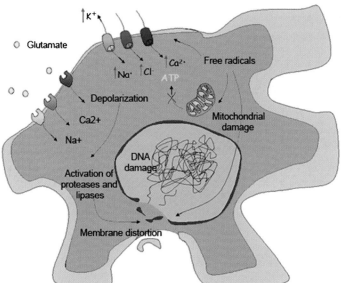

Fig. 30a–b Effects of ischemia on neuronal metabolism and integrity. **a** Brain ischemia results in an alteration in the metabolic functions of the neuron that, if prolonged, may lead to cell death. Microcirculation and glia are responsible for inflammatory and macrophagic responses that can repair the cell or favor the mechanisms of apoptosis. **b** There is an increase in free oxygen radicals and an increase in membrane permeability, which imply changes in ionic concentrations and thus the electrical activity of the neuron, mitochondrial and DNA damage, and enzymatic cascades, which would all lead to apoptosis

cells (macrophages), activation of neurogenesis with an increase in neuroblast migration speed, and neovascularization (mediated by growth factor release and signaling).

When the ultrastructure of the ischemic region is examined, two morphologically different zones can be discerned: the ischemic nucleus and the penumbra zone (**Fig.31**). The former is a necrotic area containing abundant cell detritus with

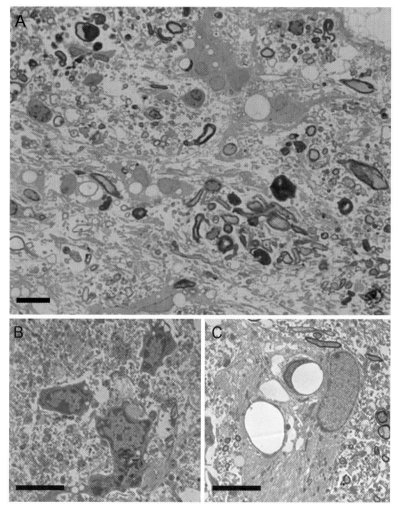

Fig.31a–c Effects of ischemia on the adult striatum. **a** Semithin section showing the ischemic nucleus in the adult striatum after focal transitory ischemia. Apoptotic nuclei, cellular detritus, oligodendrocytes, microglia, reactive astrocytes, and myelinic axons can be observed, *scale bar* 10 μm. **b** Microglial cells involved in the inflammatory response under electron microscopy, *scale bar* 10 μm. **c** Reactive astrocytes, *scale bar* 5 μm

reactive astrocytes and microglia in the close vicinity. The penumbra zone mainly comprises retracted tissue with damaged (although not dead) neurons, which mainly affect chromatin. After a month, remyelination has occurred.

It is known that in reptiles, birds and mammals, stem cells activate when lesions occur, although we are still not aware of the factors which regulate this proliferation-neurogenesis. After cerebral infarction, the brain responds by reverting back to a state similar to that of development. The neurogenesis that occurs under normal circumstances in the neuroproliferative regions of the adult mammalian brain, including the SVZ and the hippocampal DG, may increase under pathologic conditions such as cerebral infarction (Chopp and Li 2002; Gu et al. 2000; Jin et al. 2001; Liu et al. 1998) or other types of cerebral injury (Gould and Tanapat 1997; Magavi et al. 2000; Parent et al. 1997; Yoshimura et al. 2001). Several works have demonstrated the existence of neurogenesis in experimental ischemia models in mice (Arvidsson et al. 2001a, b; Jin et al. 2001; Liu et al. 1998; Thored et al. 2006; Zhang et al. 2006). After a stroke, the SVZ GFAP-expressing cells (Type B cells), which were labeled using a cell type-specific viral infection method, were found to generate neuroblasts that migrated toward the injured striatum after middle cerebral artery occlusion. SVZ-derived neuroblasts differentiated into mature neurons in the striatum, where they expressed neuronal-specific nuclear protein and formed synapses with neighboring striatal cells (Yamashita et al. 2006). These newborn neurons may compensate for the loss of neuronal function caused by strokes. Arvidsson et al. injected BrdU into rats after inducing cerebral infarction. To show the genesis of new neurons, they also used Dcx and NeuN. The number of neoformed cells detected in the striatum of the affected hemisphere was higher than in control animals. Two weeks after infarction, 20% of the striatal cells that were stained with Dcx were also stained with NeuN, thus confirming that some neuroblasts had differentiated to mature neurons. Five weeks after infarction, 42% of the BrdU- and NeuN-stained cells were also stained with DARPP-32, a marker of medium-sized spiny neurons, which represent the dominant type of striatal neuron. In other words, some neurons were differentiated to the phenotype of the neurons that had been destroyed by infarction. However, these cells were only found in the striatum, not in the damaged cortex, and 80% of the new neurons died within six weeks of infarction. This is why only some striatal neurons were replaced through neurogenesis. This data demonstrates that ischemia may lead to a proliferation of new neurons which migrate to the ischemic region, and that the markers of the mature neurons are expressed. A series of subsequent works have replicated these results.

Other stroke models also show the existence of different mechanisms which, when operated remotely, activate cellular proliferation in both the SVZ and RMS; for example, that carried out by Gotts and Chesselet (2005) in a model of cortical lesion with thermocoagulation of pial blood vessels. After provoking a lesion in the cortex, the experiment consisted of injecting 40 nL of BrdU in the left lateral ventricle of the SVZ. Animals were sacrificed after 4 h, 12 h, and 4 days. The density of the BrdU-marked cells was measured in both the SVZ and the RMS cell flow in order to show an increase in the number of BrdU$^+$ cells and to demonstrate that the lesion did not interrupt the tangential migration to the OB. Using the

TUNEL technique, which measures cellular death by apoptosis, it was proven that the increase of proliferation did not correlate with cellular death in the SVZ, which resulted in cellular accumulation. Thus, we conclude that there may be molecular signals that chemotactically drive stem cells to the lesion area.

5.3
Extracellular Factors and Neurogenesis After Stroke

The aim of the search for new therapies for brain stroke is to determine the signals that activate neurogenesis, the migration of newly formed cells, and their integration into injured areas. Some advances have been achieved in this field; it has been shown that, concomitantly to ischemic injuries, the expression of some growth factors that influence the NSCs dynamics increases (Kawahara et al. 1999; Marti 2004) (Table 5.3). Cerebral ischemia upregulates the expression of several factors: brain-derived neurotrophic factor (BDNF), glial-derived neurotrophic factor (GDNF), bFGF, insulin-like growth factor (IGF-1), erythropoietin (EPO), vascular endothelial growth factor (VEGF), and heparin-binding epidermal growth factor (HB-EGF). These factors have been shown to increase the neurogenic properties of neural stem cells. For instance, EPO is expressed in the adult brain, and in vitro and in vivo studies on neural stem cells revealed that the increased EPO gene expression was accompanied by increased neurogenesis (Shingo et al. 2001). EPO could negatively regulate the proliferation of stem cells while favoring differentiation toward a neuronal lineage. VEGF specifically regulates endothelial cell growth and differentiation and is also a survival factor for endothelial cells (Risau 1997). The observation that the VEGF receptor is expressed in neural progenitor cells (Jin et al. 2002b; Yang and Cepko 1996), and that angiogenesis and neurogenesis occur concurrently in the adult dentate gyrus (Palmer et al. 2000), suggests a potential role for VEGF in neurogenesis. In the adult murine brain, administration of exogenous VEGF increased the proliferation of neuronal precursors in both the SVZ and the DG (Jin et al. 2002b). This effect was likely mediated by the modulation of cell division rather than survival (Jin et al. 2002b). HB-EGF is a mitogenic and chemotactic glycoprotein that contains an EGF-like domain and acts through several receptors, including ErbB1, ErbB4, and heparin sulfate proteoglycans. As with EPO and VEGF, the expression of HB-EGF in the brain is increased by ischemia and results in neuroprotection (Kawahara et al. 1999). IGF-1 expression by activated astrocytes increases in the ischemic penumbra. In addition, inhibiting IGF-1 activity by intraventricular infusion of IGF-1 antibody significantly prevented ischemia-induced neural progenitor proliferation. These results indicate that the IGF-1 formed in the ischemic penumbra might be one of the diffusible factors that mediate postischemic neural progenitor proliferation.

From the data available, we deduce that there is a close relationship between angiogenesis and neurogenesis (Thored et al. 2007). The endothelial cells formed after stroke (under the effect of VEGF, EPO, nitric oxide) foster conditions leading to the local proliferation and differentiation of endogenous brain cells. This has been referred to as the "vascular niche" (Palmer et al. 2000).

Table 5.3 Neurogenesis in experimental models of cerebral ischemia in rodents (modified from Zhang et al. 2005)

Model	Factor	Neurogenesis	References
Global ischemia		Dentate gyrus	Liu et al. (1998), Schmidt and Reymann (2002), Iwai et al. (2002), Tanaka et al. (2004), Kee et al. (2001), and Yagita et al. (2001)
	bFGF	Dentate gyrus SVZ Cerebral cortex	Matsuoka et al. (2003) Nakatomi et al. (2002)
Focal ischemia		SVZ	Jiang et al. (2001)
		Dentate gyrus	Arvidsson et al. (2002)
		Cerebral cortex	Zhang et al. (2004) Jin et al. (2001)
	bFGF	Dentate gyrus	Yoshimura et al. (2001)
	G-CSF	SVZ	Shyu et al. (2004)
	EGF	SVZ	Teramoto et al. (2003)
	HB-EGF	Dentate gyrus SVZ	Jin et al. (2004)
	EPO	SVZ	Wang et al. (2004)
	Statins	SVZ Dentate gyrus	Chen et al. (2003)
	Sildenafil	SVZ Dentate gyrus	Zhang et al. (2002)
	VEGF	SVZ Dentate gyrus	Jin et al. (2002b)
	CD34$^+$ cells	SVZ	Taguchi et al. (2004)
	iNOS	Dentate gyrus	Zhu et al. (2003)

5.4
Stem Cell-Based Therapies in Ischemia

The aim of the treatment of stroke is to limit cerebral lesion. Possible strategies for treating ischemic injuries include (a) restoration of blood flow, (b) neuroprotection (preventing damaged neurons from undergoing apoptosis in the acute phase of cerebral ischemia), and (c) neurosupplementation (the repair of broken neuronal networks with newly born neurons in the chronic phase of cerebral ischemia).

Treatment of stroke has emphasized prevention and protection, but little has been done in the brain to repair stroke. Up until the last decade, the only way to tackle this problem was by means of medicines and neuroprotection. When cerebral damage occurs, the blood flow begins to diminish, producing metabolism and ionic changes as well as metabolic energy changes. Using neuroprotection, attempts have been made to discover the stages of the ischemic cascade while avoiding cytotoxicity in cerebral ischemia. Neuroprotective drugs work to minimize the effects of the ischemic cascade. While no neuroprotective agents are currently available commercially, several different types of these drugs are presently undergoing clinical trials for acute ischemic stroke. Some types of neuroprotective drugs currently being researched include glutamate antagonists, calcium antagonists, opiate antagonists, and antioxidants such as GABA-A agonist.

An approach to functional repair would be to replace the damaged tissue with new cells or to enhance endogenous neurogenesis. The new way to limit the damaged area is by means of "cellular therapy," the aims of which are to reduce the lesion area and to help the functional recovery of the damaged area through the survival, migration, and differentiation of stem cells. Using cellular therapy, it has been possible to increase migration and even partial differentiation in the grafted stem cells (Lindvall and Kokaia 2006). Enhanced neurogenesis in the SVZ area was observed after ischemic lesion in the striatum. These immature neurons migrate into the damaged striatum, where they express the markers of striatal medium spiny projection neurons. Thus, the new neurons seem to differentiate into the phenotype of most of the neurons destroyed by ischemic lesion. However, as >80% of the new neurons die in the first few weeks following a stroke, they only replace a small fraction (0.2%) of the mature striatal neurons that have died.

The precursor cells must be connected with the appropriate neurons of the appropriate layer. Restitution of function probably requires not only new neurons but also reconstruction of the medium, including the glia, the astrocytes, the angioblasts, and the extracellular matrix. Basically, an environment with the following features should be achieved: proangiogenic, anti-inflammatory, antiexcitotoxic, and antiapoptotic.

At the molecular level, genes and proteins that are common in embryonic development are expressed, such as nestin, neuroD, growth-associated protein 43, synaptophysine, and the aforementioned growth factors, which promote neuronal development, synaptogenesis, and angiogenesis. These proteins are particularly abundant in the region adjacent to the ischemic lesion. It is very important to identify the trophic factors that enhance the maturation and survival of immature neurons, as this could provide important clues about how to improve recovery after neonatal brain injury (Ong et al. 2005). These factors could act by cooperating with endogenous neurogenesis, or could improve the results of cell transplantation. It seems that a combinational therapy of factors or their inhibitors may provide powerful therapeutic potential for enhancing stroke-induced neurogenesis and restoring damaged tissue in order for it to function once more.

6
Therapeutic Potential of Neural Stem Cells

Stem cell therapy has become a promising prospect in the treatment of degenerative diseases of different body organs and systems. In this context, both embryonic stem cells and adult stem cells have been considered for transplants in these cell-based therapies (Jordan et al. 2006). The cellular behavior of NSCs (or neural progenitors) has allowed us to keep them in culture, and to amplify their number after serial passages. This implies that adult NSCs may be an important source of cells for transplants.

As a first step to studying the therapeutic applications of stem cells, it is crucial to demonstrate that the implanted cells are able to graft into the tissue over both the short term and the long term (without tumoral degeneration) so that their migratory capacity and their potential to become functionally integrated can be determined. When these aspects have been demonstrated, certain studies of experimental models of diseases must be carried out to show that implanted cells guarantee clinical improvement and prolonged survival in the patient. Once these requirements have been met, the cells can be explored in clinical assays. The fact that stem cells can be useful in therapies utilizing different clinical approaches must be considered. NSCs can be used as replacements when neurons are lost as a result of an injury (stroke, trauma, or degenerative disorders) (Jordan et al. 2006). This requires a cell niche, with the proper environmental cues, where cells would be likely to differentiate into the appropriate cell type, integrate, and become functional. The surrounding inflammatory process, which is secondary to the lesion in most pathologic processes, could become an obstacle for these cells. Microglia, astrocytes, and peripheral macrophages are the key players mediating this response, as they secrete an array of pro- and anti-inflammatory cytokines and chemokines. On the one hand, these mediators act by conferring immune protection to the brain; on the other, they lead to neuronal death, causing a major alteration in the "neurogenic niche" (Basu et al. 2005; Taupin 2006). An ideal pharmacological approach would be to modulate the inflammatory response in the desired proneurogenic way.

With regards to the preferential use of embryonic stem cells or adult stem cells, both have been exposed to a certain degree of criticism due to their limitations, which need to be reconciled before they can be used in clinical research (Table 6.1). Embryonic stem cells (ESCs) can cause tumors, and may be rejected by the immune system. A significant proportion of society also questions their use from an ethical

Table 6.1 Adult neural stem cells versus embryonic stem cells in human therapies

ESCs		aNSCs	
Advantages	Criticism	Advantages	Criticism
Indefinite self-renewal	Tumors	No rejection	Restricted differentiation potential
Multipotency	Immune rejection	No tumors	Small quantities
Availability	Ethics		Decreased proliferative potential

perspective. In turn, while ESCs can self-renew indefinitely and can form many different cell types, the full potential of adult stem cells is uncertain; in fact, there is evidence to suggest that they may be more limited. Conversely, adult cells incur no risk of rejection because they can originate from the patients' own bodies (autologous transplantation) and they do not appear to cause tumors. However, they may be present in small quantities (Nunes et al. 2003; Roy et al. 2000b) and are very difficult to isolate. Most of the time, they remain in regions that are not easily accessible. For example, adult NSCs reside deep in the telencephalon, surrounded by essential structures; any damage caused to these structures could be very risky for patients. Another limitation of adult stem cells is that they may only be able to divide a limited number of times (Kim and Morshead 2003), which would limit their usefulness in the production of adequate numbers of well-characterized cells for reliable therapies.

The administration protocol is susceptible to change depending on whether a multifocal disease (where systemic administration has been demonstrated to be more efficacious) (Pluchino et al. 2003) or focal injury (where intralesional injection facilitates tissue regeneration) is being addressed. Another possible strategy is the exogenous "activation" of endogenous NSCs, i.e., those that already reside in the stem cell niche, to proliferate and migrate toward the lesion, where they become new, fully operational neurons (Chmielnicki et al. 2004). It is clear that "intrinsic" self-repair capabilities are not enough to promote complete and lasting CNS repair. Noninvasive methods exist that increase the number of adult stem cells: exercise, hormones, calorie restriction, antioxidants, etc., as reviewed in Lie et al. (2004). Growth factors, and possibly their small-molecule mimics, can also alter the cell proliferation rate and neurogenesis, as mentioned in Chap. 5 for ischemic brain injury.

An unexpected method of action has been demonstrated in NSC therapies and has been termed the "bystander effect," as reviewed by Martino and Pluchino (Martino and Pluchino 2006; Pluchino et al. 2007). This concept derives from studies in which the transplantation of NSCs significantly improved the function of the target organs. However, histological studies have never demonstrated an increase in neuronal incorporation or cell-based repair. Therefore, this effect may be the result of the enhanced survival of existing neurons, or delayed death, by means of the stem cell-dependent delivery of trophic factors or other molecules to the target injured site. Prosurvival factors would exist in the injured region at sufficient levels without producing side effects in other brain regions. To achieve this, NSCs must be able to survive in the niche for sufficiently long periods to allow them to release factors and reduce the logical and negative outcomes of the disease.

Another approach in stem cell therapy concerns cell fusion. Cell fusion is an old phenomenon, and one that is especially relevant in infectious processes, but it is also a main functional mechanism of body tissues, e.g., for the fibers of the skeletal muscle or osteoclastic cells. In this sense, we should not forget that life starts with the fusion of two germinal cells. Until recently, it was thought that multinucleated cells were formed from nuclei of the same cell type, but a new research field has recently investigated the possibility of the fusion of cells from different germinal lines, therefore nuclei from different origins coexist in the same cell type. Given their stem cell potential, their presence in the peripheral blood and their plasticity, bone marrow stem cells (BMSCs) are considered to be the cells that fuse with static cells throughout the organism (Lagasse et al. 2000). During the fusion process, BMSCs provide new genetic material to the resident cell and provided it with a new identity. The tissue regeneration processes and the data obtained and published from bone marrow transplantation could be explained completely based on only the cell fusion theory.

Terada and colleagues (2002) were the first researchers to show cell fusion between bone marrow cells and embryonic stem cells cocultured in vitro. Each cell type possessed a resistance gene for an antibiotic. When both antibiotics were added to the medium, some cells grew because they were doubly resistant. These results were confirmed by performing a karyotype analysis on the fused cells, which showed 4n DNA content and new phenotypic features. This discovery was important because it demonstrated the possibility of reprogramming genes of a senescent cell to induce their proliferation after cell fusion (Rodic et al. 2004). Apparently, cell fusion is a mechanism that can rescue damaged cells from sure death. It has now been demonstrated that liver regeneration is the consequence of cell fusion with BMSCs, particularly the monocytic type of BMSCs (Vassilopoulos et al. 2003; Wang et al. 2003). It is important to note that other cell types beyond liver cells are capable of fusing with BMSCs. This is the case for the Purkinje neuronal cells in the cerebellum or cardiac muscle fibers (Alvarez-Dolado et al. 2003). We still do not know whether fused cells proliferate, even when unpublished results from our group confirm they do. It is also unclear whether the extra genetic material is expressed or whether the nuclei from resident cells regenerate after damage or degenerate.

A detailed understanding of the processes and factors influencing neurogenesis would provide us with sufficient knowledge to alter the pattern of new neuron formation and to unravel the limited regenerative capacities of endogenous stem cells after injury. In any case, the potential for stem cell therapies in neurodegenerative diseases remains encouraging, although important questions must be answered before we can predict which strategy (type and age of cells, administration procedure, pharmacological support, etc.) will be most appropriate for which disease.

7
Concluding Remarks

The concept of adult neurogenesis is relatively modern. It surprised the scientific community and ruled out the previously established idea that we are born with a set number of neurons. This breakthrough occurred due to various advances made throughout the last century, including morphological studies, the use of markers for proliferation, the discovery of molecular markers for specific cell types, and the accuracy of electron microscopy, among others. None of these advances was sufficient alone, but when they were all combined in elegant assays, it was possible to confirm the existence of neurogenesis throughout adulthood. It was initially thought to be only restricted to reptiles and birds, but it has now also been demonstrated in mammals, including humans.

There are two well-accepted neurogenic regions in the adult mammalian brain: the olfactory bulb and the hippocampus. Stem cells with neurogenic potential have also been found in the third ventricle, in the central canal of the spinal cord, and recently in the subcallosal region. The majority of the new neurons that arrive in the OB are generated in an area that underlies the lateral wall of the lateral ventricles: the SVZ. There are different cell populations in the SVZ that have been widely described using EM. Type B cells are astrocytic in nature and have been identified as adult neural stem cells. Type C cells are transit-amplifying precursors derived from the Type B cells that proliferate at a high rate and give rise to Type A cells. Type A cells express the markers of immature neurons and are able to migrate toward the OB. This migration occurs along a very well-characterized path called the RMS in which structural astrocytes surround the population of neuroblasts to create a permissive environment for migration. Type E cells correspond to ependymal cells that do not divide in any case and play a structural and protective role, as well as probably acting in cell migration. Type E cells are controversial, as some groups affirm that primordial neural stem cells are ependymal cells. This remains a matter of debate. The DG of the hippocampus has also been studied for its proliferation and neurogenic potential. It has been proposed to be a source of adult NSCs that migrate within the hippocampus to the granular layer and become new functional neurons involved in memory processes.

Specific molecular markers have helped to identify different cell types in the SVZ. By combining them with markers of proliferation, they have proven useful for recognizing which of them undergo proliferation and what their rate of division is from specific lineages. Based on the specificity of surface markers, and by using FACS, we have been able to purify these cell populations for further studies. In morphological terms, EM has also proven essential in the identification of the cell types involved in adult neurogenesis. Time-dependent studies with 3HT or BrdU with LM have elucidated the temporal profile and lineage progression, ranging from the generation of new cells to integration into the OB. Furthermore, the neurosphere assay has also been essential in the characterization of adult NSCs. The neurosphere assay is the result of culturing cells derived from the adult SVZ. This results in the generation of clonal growths or neurospheres, which are representative of the neural stem cell population and neuronal precursors. However, given its multiple applications in experimental designs, in amplifying cell populations, or in cell transplantation, we must be very cautious when extrapolating the results of the neurosphere assay performed in vitro to the in vivo behavior of stem cells. Clonal neurospheres are composed not only of adult NSCs but of a mixture of proliferating and differentiated cells with a low degree of stemness and more restricted potential.

Recently, new molecular strategies and the use of transgenic animals have provided us with an insight into the gain or loss of function of different genes that may be involved in proliferation or neurogenesis. The cre–lox system allows us to conditionally express or knock down certain genes, and also to study the phenomenon of cell fusion. In vivo electrophysiology of labeled cells has demonstrated that adult NSCs from the SVZ integrate, are functional in the OB, and are capable of triggering electric impulses.

All of these studies have been performed in the mouse SVZ, but neurogenesis has been demonstrated in other mammals such as rabbits, cows, monkeys, and humans. This is very important, as the meaning of neurogenesis in humans remains unknown, and it is essential to gain an understanding of it in order to effectively apply it in therapies for neurodegenerative diseases. There are striking differences between the mouse and human SVZ, not only in the cytoarchitecture of the SVZ layers, but also in the cell types and their behavior. While the existence of immature neuroblasts has been demonstrated in the human SVZ, it is not clear whether the RMS persists in the human brain, at least as it is described in rodents. However, a recent paper describes chain migration in the adult human brain. This contradiction might be due to the difficulty involved in obtaining human samples under adequate conditions for histologic studies.

Adult neurogenesis is a process that under physiological conditions is not assumed to be highly significant. When injured, neurogenic regions are activated, leading to increased proliferation and neurogenic potential. This is observed in response to ischemia. Stroke has become an important target of stem cell-based therapies as the social repercussions of stroke have emerged in recent years. The potential of adult NSCs to populate ischemic areas is being studied, and these investigations have produced promising results. Stem cell therapy for a number of neurological diseases is becoming a feasible possibility. There are different approaches to treatments with stem cells, including exogenous cell administration or enhancement of endogenous neurogenesis.

Another perspective to consider is the potential oncogenicity that undifferentiated cells in the adult brain may retain. These cells are quiescent but can suffer changes or mutations in response to external or internal cues that might lead to tumor formation. Many scientific works have followed this line, and some advances are shedding light on the origin and treatment of brain tumors. This topic is also important because stem cell-based therapies could graft potential neoplasms into the brain instead of curing a disease; this is more likely for embryonic stem cells.

In recent decades, the discovery of the existence of adult neurogenesis has led to the reopening of a field that had been closed since the start of the twentieth century. We continue to advance in neural stem cell research due to new technical approaches like genomics and proteomics. Previous knowledge and new discoveries will hopefully shed some light on as-yet unknown aspects of the neurogenic process, thus making the use of adult NSCs in human therapy safe.

References

Aboody KS, Brown A, Rainov NG, Bower KA, Liu S, Yang W, Small JE, Herrlinger U, Ourednik V, Black PM, Breakefield XO, Snyder EY (2000) Neural stem cells display extensive tropism for pathology in adult brain: evidence from intracranial gliomas. Proc Natl Acad Sci USA 97(23):12846–12851

Adrian EK Jr, Walker BE (1962) Incorporation of thymidine-H3 by cells in normal and injured mouse spinal cord. J Neuropathol Exp Neurol 21:597–609

Altman J (1962) Are new neurons formed in the brains of adult mammals? Science 135:1127–1128

Altman J (1969a) Autoradiographic and histological studies of postnatal neurogenesis. 3. Dating the time of production and onset of differentiation of cerebellar microneurons in rats. J Comp Neurol 136(3):269–293

Altman J (1969b) Autoradiographic and histological studies of postnatal neurogenesis. IV. Cell proliferation and migration in the anterior forebrain, with special reference to persisting neurogenesis in the olfactory bulb. J Comp Neurol 137(4):433–457

Altman J, Bayer SA (1979a) Development of the diencephalon in the rat. IV. Quantitative study of the time of origin of neurons and the internuclear chronological gradients in the thalamus. J Comp Neurol 188(3):455–471

Altman J, Bayer SA (1979b) Development of the diencephalon in the rat. V. Thymidine-radiographic observations on internuclear and intranuclear gradients in the thalamus. J Comp Neurol 188(3):473–499

Altman J, Bayer SA (1979c) Development of the diencephalon in the rat. VI. Re-evaluation of the embryonic development of the thalamus on the basis of thymidine-radiographic datings. J Comp Neurol 188(3):501–524

Altman J, Bayer SA (1981a) Time of origin of neurons of the rat inferior colliculus and the relations between cytogenesis and tonotopic order in the auditory pathway. Exp Brain Res 42(3–4):411–423

Altman J, Bayer SA (1981b) Time of origin of neurons of the rat superior colliculus in relation to other components of the visual and visuomotor pathways. Exp Brain Res 42(3–4):424–434

Altman J, Das GD (1965) Autoradiographic and histological evidence of postnatal hippocampal neurogenesis in rats. J Comp Neurol 124(3):319–335

Alvarez-Buylla A, Lim DA (2004) For the long run: maintaining germinal niches in the adult brain. Neuron 41(5):683–686

Alvarez-Buylla A, Nottebohm F (1988) Migration of young neurons in adult avian brain. Nature 335(6188):353–354

Alvarez-Dolado M, Pardal R, Garcia-Verdugo JM, Fike JR, Lee HO, Pfeffer K, Lois C, Morrison SJ, Alvarez-Buylla A (2003) Fusion of bone-marrow-derived cells with Purkinje neurons, cardiomyocytes and hepatocytes. Nature 425(6961):968–973

Arnhold S, Hilgers M, Lenartz D, Semkova I, Kochanek S, Voges J, Andressen C, Addicks K (2003) Neural precursor cells as carriers for a gene therapeutical approach in tumor therapy. Cell Transplant 12(8):827–837

Arvidsson A, Kokaia Z, Airaksinen MS, Saarma M, Lindvall O (2001a) Stroke induces widespread changes of gene expression for glial cell line-derived neurotrophic factor family receptors in the adult rat brain. Neuroscience 106(1):27–41

Arvidsson A, Kokaia Z, Lindvall O (2001b) N-methyl-D-aspartate receptor-mediated increase of neurogenesis in adult rat dentate gyrus following stroke. Eur J Neurosci 14(1):10–18

Arvidsson A, Collin T, Kirik D, Kokaia Z, Lindvall O (2002) Neuronal replacement from endogenous precursors in the adult brain after stroke. Nat Med 8(9):963–970

Asano K, Kubo O, Tajika Y, Huang MC, Takakura K, Ebina K, Suzuki S (1997) Expression and role of cadherins in astrocytic tumors. Brain Tumor Pathol 14(1):27–33

Asano K, Duntsch CD, Zhou Q, Weimar JD, Bordelon D, Robertson JH, Pourmotabbed T (2004) Correlation of N-cadherin expression in high grade gliomas with tissue invasion. J Neurooncol 70(1):3–15

Assanah M, Lochhead R, Ogden A, Bruce J, Goldman J, Canoll P (2006) Glial progenitors in adult white matter are driven to form malignant gliomas by platelet-derived growth factor-expressing retroviruses. J Neurosci 26(25):6781–6790

Bachoo RM, Maher EA, Ligon KL, Sharpless NE, Chan SS, You MJ, Tang Y, DeFrances J, Stover E, Weissleder R, Rowitch DH, Louis DN, DePinho RA (2002) Epidermal growth factor receptor and Ink4a/Arf: convergent mechanisms governing terminal differentiation and transformation along the neural stem cell to astrocyte axis. Cancer Cell 1(3):269–277

Bajenaru ML, Donahoe J, Corral T, Reilly KM, Brophy S, Pellicer A, Gutmann DH (2001) Neurofibromatosis 1 (NF1) heterozygosity results in a cell-autonomous growth advantage for astrocytes. Glia 33(4):314–323

Baker SA, Baker KA, Hagg T (2004) Dopaminergic nigrostriatal projections regulate neural precursor proliferation in the adult mouse subventricular zone. Eur J Neurosci 20(2):575–579

Balentova S, Racekova E, Martoncikova M, Misurova E (2006) Cell proliferation in the adult rat rostral migratory stream following exposure to gamma irradiation. Cell Mol Neurobiol 26(7–8):1129–1137

Banasr M, Hery M, Printemps R, Daszuta A (2004) Serotonin-induced increases in adult cell proliferation and neurogenesis are mediated through different and common 5-HT receptor subtypes in the dentate gyrus and the subventricular zone. Neuropsychopharmacology 29(3):450–460

Bao S, Wu Q, Sathornsumetee S, Hao Y, Li Z, Hjelmeland AB, Shi Q, McLendon RE, Bigner DD, Rich JN (2006) Stem cell-like glioma cells promote tumor angiogenesis through vascular endothelial growth factor. Cancer Res 66(16):7843–7848

Barami K, Iversen K, Furneaux H, Goldman SA (1995) Hu protein as an early marker of neuronal phenotypic differentiation by subependymal zone cells of the adult songbird forebrain. J Neurobiol 28(1):82–101

Barami K, Lewis-Tuffin L, Anastasiadis PZ (2006) The role of cadherins and catenins in gliomagenesis. Neurosurg Focus 21(4):E13

Barnea A, Nottebohm F (1994) Seasonal recruitment of hippocampal neurons in adult free-ranging black-capped chickadees. Proc Natl Acad Sci USA 91(23):11217–11221

Basu A, Lazovic J, Krady JK, Mauger DT, Rothstein RP, Smith MB, Levison SW (2005) Interleukin-1 and the interleukin-1 type 1 receptor are essential for the progressive neurodegeneration that ensues subsequent to a mild hypoxic/ischemic injury. J Cereb Blood Flow Metab 25(1):17–29

Bayer SA (1980a) Development of the hippocampal region in the rat. I. Neurogenesis examined with 3H-thymidine autoradiography. J Comp Neurol 190(1):87–114

Bayer SA (1980b) Quantitative 3H-thymidine radiographic analyses of neurogenesis in the rat amygdala. J Comp Neurol 194(4):845–875

Bayer SA (1983) 3H-thymidine-radiographic studies of neurogenesis in the rat olfactory bulb. Exp Brain Res 50(2–3):329–340

Beattie MS, Bresnahan JC, Komon J, Tovar CA, Van Meter M, Anderson DK, Faden AI, Hsu CY, Noble LJ, Salzman S, Young W (1997) Endogenous repair after spinal cord contusion injuries in the rat. Exp Neurol 148(2):453–463

Bernier PJ, Bedard A, Vinet J, Levesque M, Parent A (2002) Newly generated neurons in the amygdala and adjoining cortex of adult primates. Proc Natl Acad Sci USA 99(17):11464–11469

Bjornson CR, Rietze RL, Reynolds BA, Magli MC, Vescovi AL (1999) Turning brain into blood: a hematopoietic fate adopted by adult neural stem cells in vivo. Science 283(5401):534–537

Blakemore WF (1969) The ultrastructure of the subependymal plate in the rat. J Anat 104(Pt 3):423–433

Blakemore WF, Jolly RD (1972) The subependymal plate and associated ependyma in the dog. An ultrastructural study. J Neurocytol 1(1):69–84

Bonita R, Beaglehole R (1993) Explaining stroke mortality trends. Lancet 341(8859):1510–1511

Bosch FX, Udvarhelyi N, Venter E, Herold-Mende C, Schuhmann A, Maier H, Weidauer H, Born AI (1993) Expression of the histone H3 gene in benign, semi-malignant and malignant lesions of the head and neck: a reliable proliferation marker. Eur J Cancer 29A(10):1454–1461

Bravo R, Frank R, Blundell PA, Macdonald-Bravo H (1987) Cyclin/PCNA is the auxiliary protein of DNA polymerase-delta. Nature 326(6112):515–517

Brazel CY, Limke TL, Osborne JK, Miura T, Cai J, Pevny L, Rao MS (2005) Sox2 expression defines a heterogeneous population of neurosphere-forming cells in the adult murine brain. Aging Cell 4(4):197–207

Broderick JP, Phillips SJ, Whisnant JP, O'Fallon WM, Bergstralh EJ (1989) Incidence rates of stroke in the eighties: the end of the decline in stroke? Stroke 20(5):577–582

Bryans WA (1959) Mitotic activity in the brain of the adult rat. Anat Rec 133:65–73

Burns TC, Ortiz-Gonzalez XR, Gutierrez-Perez M, Keene CD, Sharda R, Demorest ZL, Jiang Y, Nelson-Holte M, Soriano M, Nakagawa Y, Luquin MR, Garcia-Verdugo JM, Prosper F, Low WC, Verfaillie CM (2006) Thymidine analogs are transferred from prelabeled donor to host cells in the central nervous system after transplantation: a word of caution. Stem Cells 24(4):1121–1127

Cameron HA, Woolley CS, McEwen BS, Gould E (1993) Differentiation of newly born neurons and glia in the dentate gyrus of the adult rat. Neuroscience 56(2):337–344

Carleton A, Petreanu LT, Lansford R, Alvarez-Buylla A, Lledo PM (2003) Becoming a new neuron in the adult olfactory bulb. Nat Neurosci 6(5):507–518

Carmichael ST (2003) Gene expression changes after focal stroke, traumatic brain and spinal cord injuries. Curr Opin Neurol 16(6):699–704

Chambers CB, Peng Y, Nguyen H, Gaiano N, Fishell G, Nye JS (2001) Spatiotemporal selectivity of response to Notch1 signals in mammalian forebrain precursors. Development 128(5):689–702

Charytoniuk D, Porcel B, Rodriguez Gomez J, Faure H, Ruat M, Traiffort E (2002) Sonic Hedgehog signalling in the developing and adult brain. J Physiol Paris 96(1–2):9–16

Chen J, Zhang ZG, Li Y, Wang Y, Wang L, Jiang H, Zhang C, Lu M, Katakowski M, Feldkamp CS, Chopp M (2003) Statins induce angiogenesis, neurogenesis, and synaptogenesis after stroke. Ann Neurol 53(6):743–751

Cheng L, Fu J, Tsukamoto A, Hawley RG (1996) Use of green fluorescent protein variants to monitor gene transfer and expression in mammalian cells. Nat Biotechnol 14(5):606–609

Chmielnicki E, Benraiss A, Economides AN, Goldman SA (2004) Adenovirally expressed noggin and brain-derived neurotrophic factor cooperate to induce new medium spiny neurons from resident progenitor cells in the adult striatal ventricular zone. J Neurosci 24(9):2133–2142

Chopp M, Li Y (2002) Treatment of neural injury with marrow stromal cells. Lancet Neurol 1(2):92–100

Chou MY, Chang AL, McBride J, Donoff B, Gallagher GT, Wong DT (1990) A rapid method to determine proliferation patterns of normal and malignant tissues by H3 mRNA in situ hybridization. Am J Pathol 136(4):729–733

Clarke DL, Johansson CB, Wilbertz J, Veress B, Nilsson E, Karlstrom H, Lendahl U, Frisen J (2000) Generalized potential of adult neural stem cells. Science 288(5471):1660–1663

Conover JC, Doetsch F, Garcia-Verdugo JM, Gale NW, Yancopoulos GD, Alvarez-Buylla A (2000) Disruption of Eph/ephrin signaling affects migration and proliferation in the adult subventricular zone. Nat Neurosci 3(11):1091–1097

Cooper O, Isacson O (2004) Intrastriatal transforming growth factor alpha delivery to a model of Parkinson's disease induces proliferation and migration of endogenous adult neural progenitor cells without differentiation into dopaminergic neurons. J Neurosci 24(41):8924–8931

Coronas V, Bantubungi K, Fombonne J, Krantic S, Schiffmann SN, Roger M (2004) Dopamine D3 receptor stimulation promotes the proliferation of cells derived from the post-natal subventricular zone. J Neurochem 91(6):1292–1301

Coskun V, Venkatraman G, Yang H, Rao MS, Luskin MB (2001) Retroviral manipulation of the expression of bone morphogenetic protein receptor Ia by SVZa progenitor cells leads to changes in their p19(INK4d) expression but not in their neuronal commitment. Int J Dev Neurosci 19(2):219–227

Coskun V, Wu H, Blanchi B, Tsao S, Kim K, Zhao J, Biancotti JC, Hutnick L, Krueger RC Jr, Fan G, de Vellis J, Sun YE (2008) CD133+ neural stem cells in the ependyma of mammalian postnatal forebrain. Proc Natl Acad Sci USA 105(3):1026–1031

Crews FT, Miller MW, Ma W, Nixon K, Zawada WM, Zakhari S (2003) Neural stem cells and alcohol. Alcohol Clin Exp Res 27(2):324–335

Curtis MA, Penney EB, Pearson AG, van Roon-Mom WM, Butterworth NJ, Dragunow M, Connor B, Faull RL (2003) Increased cell proliferation and neurogenesis in the adult human Huntington's disease brain. Proc Natl Acad Sci USA 100(15):9023–9027

Curtis MA, Kam M, Nannmark U, Anderson MF, Axell MZ, Wikkelso C, Holtas S, van Roon-Mom WM, Bjork-Eriksson T, Nordborg C, Frisen J, Dragunow M, Faull RL, Eriksson PS (2007) Human neuroblasts migrate to the olfactory bulb via a lateral ventricular extension. Science 315(5816):1243–1249

Dahlstrand J, Collins VP, Lendahl U (1992) Expression of the class VI intermediate filament nestin in human central nervous system tumors. Cancer Res 52(19):5334–5341

Dai C, Celestino JC, Okada Y, Louis DN, Fuller GN, Holland EC (2001) PDGF autocrine stimulation dedifferentiates cultured astrocytes and induces oligodendrogliomas and oligoastrocytomas from neural progenitors and astrocytes in vivo. Genes Dev 15(15):1913–1925

Daou MC, Smith TW, Litofsky NS, Hsieh CC, Ross AH (2005) Doublecortin is preferentially expressed in invasive human brain tumors. Acta Neuropathol 110(5):472–480

Davis EE, Brueckner M, Katsanis N (2006) The emerging complexity of the vertebrate cilium: new functional roles for an ancient organelle. Dev Cell 11(1):9–19

Dayer AG, Cleaver KM, Abouantoun T, Cameron HA (2005) New GABAergic interneurons in the adult neocortex and striatum are generated from different precursors. J Cell Biol 168(3):415–427

Dervan AG, Roberts BL (2003a) Reaction of spinal cord central canal cells to cord transection and their contribution to cord regeneration. J Comp Neurol 458(3):293–306

Dervan AG, Roberts BL (2003b) The meningeal sheath of the regenerating spinal cord of the eel, Anguilla. Anat Embryol 207(2):157–167

Doetsch F, Garcia-Verdugo JM, Alvarez-Buylla A (1997) Cellular composition and three-dimensional organization of the subventricular germinal zone in the adult mammalian brain. J Neurosci 17(13):5046–5061

Doetsch F, Caille I, Lim DA, Garcia-Verdugo JM, Alvarez-Buylla A (1999) Subventricular zone astrocytes are neural stem cells in the adult mammalian brain. Cell 97(6):703–716

Doetsch F, Petreanu L, Caille I, Garcia-Verdugo JM, Alvarez-Buylla A (2002) EGF converts transit-amplifying neurogenic precursors in the adult brain into multipotent stem cells. Neuron 36(6):1021–1034

Dolbeare F (1995) Bromodeoxyuridine: a diagnostic tool in biology and medicine, Part I: historical perspectives, histochemical methods and cell kinetics. Histochem J 27(5):339–369

Dolbeare F, Selden JR (1994) Immunochemical quantitation of bromodeoxyuridine: application to cell-cycle kinetics. Methods Cell Biol 41:297–316

Dore-Duffy P, Katychev A, Wang X, Van Buren E (2006) CNS microvascular pericytes exhibit multipotential stem cell activity. J Cereb Blood Flow Metab 26:613–624

Draetta G, Beach D (1989) The mammalian cdc2 protein kinase: mechanisms of regulation during the cell cycle. J Cell Sci Suppl 12:21–27

Draetta G, Brizuela L, Beach D (1988) p34, a protein kinase involved in cell cycle regulation in eukaryotic cells. Adv Exp Med Biol 231:453–457

Eckenhoff MF, Rakic P (1991) A quantitative analysis of synaptogenesis in the molecular layer of the dentate gyrus in the rhesus monkey. Brain Res Dev Brain Res 64(1–2):129–135

Ellis P, Fagan BM, Magness ST, Hutton S, Taranova O, Hayashi S, McMahon A, Rao M, Pevny L (2004) SOX2, a persistent marker for multipotential neural stem cells derived from embryonic stem cells, the embryo or the adult. Dev Neurosci 26(2–4):148–165

Emsley JG, Hagg T (2003) Endogenous and exogenous ciliary neurotrophic factor enhances forebrain neurogenesis in adult mice. Exp Neurol 183(2):298–310

Episkopou V (2005) SOX2 functions in adult neural stem cells. Trends Neurosci 28(5):219–221

Eriksson PS, Perfilieva E, Bjork-Eriksson T, Alborn AM, Nordborg C, Peterson DA, Gage FH (1998) Neurogenesis in the adult human hippocampus. Nat Med 4(11):1313–1317

Ferri AL, Cavallaro M, Braida D, Di Cristofano A, Canta A, Vezzani A, Ottolenghi S, Pandolfi PP, Sala M, DeBiasi S, Nicolis SK (2004) Sox2 deficiency causes neurodegeneration and impaired neurogenesis in the adult mouse brain. Development 131(15):3805–3819

Font E, Desfilis E, Perez-Canellas M, Alcantara S, Garcia-Verdugo JM (1997) 3-Acetylpyridine-induced degeneration and regeneration in the adult lizard brain: a qualitative and quantitative analysis. Brain Res 754(1–2):245–259

Forbes JF (1993) Cost of stroke. Scott Med J 38(3 Suppl):S4–S5

Frisen J, Johansson CB, Lothian C, Lendahl U (1998) Central nervous system stem cells in the embryo and adult. Cell Mol Life Sci 54(9):935–945

Gabay L, Lowell S, Rubin LL, Anderson DJ (2003) Deregulation of dorsoventral patterning by FGF confers trilineage differentiation capacity on CNS stem cells in vitro. Neuron 40(3):485–499

Galli R, Borello U, Gritti A, Minasi MG, Bjornson C, Coletta M, Mora M, De Angelis MG, Fiocco R, Cossu G, Vescovi AL (2000a) Skeletal myogenic potential of human and mouse neural stem cells. Nat Neurosci 3(10):986–991

Galli R, Pagano SF, Gritti A, Vescovi AL (2000b) Regulation of neuronal differentiation in human CNS stem cell progeny by leukemia inhibitory factor. Dev Neurosci 22(1–2):86–95

Garcia-Verdugo JM, Farinas I, Molowny A, Lopez-Garcia C (1986) Ultrastructure of putative migrating cells in the cerebral cortex of Lacerta galloti. J Morphol 189(2):189–197

Garcia-Verdugo JM, Doetsch F, Wichterle H, Lim DA, Alvarez-Buylla A (1998) Architecture and cell types of the adult subventricular zone: in search of the stem cells. J Neurobiol 36(2):234–248

Garcia-Verdugo JM, Ferron S, Flames N, Collado L, Desfilis E, Font E (2002) The proliferative ventricular zone in adult vertebrates: a comparative study using reptiles, birds, and mammals. Brain Res Bull 57(6):765–775

Gerdes J, Schwab U, Lemke H, Stein H (1983) Production of a mouse monoclonal antibody reactive with a human nuclear antigen associated with cell proliferation. Int J Cancer 31(1):13–20

Gerdes J, Lemke H, Baisch H, Wacker HH, Schwab U, Stein H (1984) Cell cycle analysis of a cell proliferation-associated human nuclear antigen defined by the monoclonal antibody Ki-67. J Immunol 133(4):1710–1715

Gil-Perotin S, Marin-Husstege M, Li J, Soriano-Navarro M, Zindy F, Roussel MF, Garcia-Verdugo JM, Casaccia-Bonnefil P (2006) Loss of p53 induces changes in the behavior of subventricular zone cells: implication for the genesis of glial tumors. J Neurosci 26(4):1107–1116

Glass R, Synowitz M, Kronenberg G, Walzlein JH, Markovic DS, Wang LP, Gast D, Kiwit J, Kempermann G, Kettenmann H (2005) Glioblastoma-induced attraction of endogenous neural precursor cells is associated with improved survival. J Neurosci 25(10):2637–2646

Goldman SA, Nottebohm F (1983) Neuronal production, migration, and differentiation in a vocal control nucleus of the adult female canary brain. Proc Natl Acad Sci USA 80(8):2390–2394

Gotts JE, Chesselet MF (2005) Vascular changes in the subventricular zone after distal cortical lesions. Exp Neurol 194(1):139–150

Gould E (2007) How widespread is adult neurogenesis in mammals? Nat Rev 8(6):481–488

Gould E, Gross CG (2002) Neurogenesis in adult mammals: some progress and problems. J Neurosci 22(3):619–623

Gould E, Tanapat P (1997) Lesion-induced proliferation of neuronal progenitors in the dentate gyrus of the adult rat. Neuroscience 80(2):427–436

Gould E, McEwen BS, Tanapat P, Galea LA, Fuchs E (1997) Neurogenesis in the dentate gyrus of the adult tree shrew is regulated by psychosocial stress and NMDA receptor activation. J Neurosci 17(7):2492–2498

Gould E, Tanapat P, McEwen BS, Flugge G, Fuchs E (1998) Proliferation of granule cell precursors in the dentate gyrus of adult monkeys is diminished by stress. Proc Natl Acad Sci USA 95(6):3168–3171

Gould E, Reeves AJ, Fallah M, Tanapat P, Gross CG, Fuchs E (1999a) Hippocampal neurogenesis in adult Old World primates. Proc Natl Acad Sci USA 96(9):5263–5267

Gould E, Reeves AJ, Graziano MS, Gross CG (1999b) Neurogenesis in the neocortex of adult primates. Science 286(5439):548–552

Gould E, Tanapat P, Hastings NB, Shors TJ (1999c) Neurogenesis in adulthood: a possible role in learning. Trends Cogn Sci 3(5):186–192

Gould E, Tanapat P, Rydel T, Hastings N (2000) Regulation of hippocampal neurogenesis in adulthood. Biol Psychiatry 48(8):715–720

Gritti A, Bonfanti L, Doetsch F, Caille I, Alvarez-Buylla A, Lim DA, Galli R, Verdugo JM, Herrera DG, Vescovi AL (2002) Multipotent neural stem cells reside into the rostral extension and olfactory bulb of adult rodents. J Neurosci 22(2):437–445

Gross RE, Mehler MF, Mabie PC, Zang Z, Santschi L, Kessler JA (1996) Bone morphogenetic proteins promote astroglial lineage commitment by mammalian subventricular zone progenitor cells. Neuron 17(4):595–606

References

Gu W, Brannstrom T, Wester P (2000) Cortical neurogenesis in adult rats after reversible photothrombotic stroke. J Cereb Blood Flow Metab 20(8):1166–1173

Han YG, Spassky N, Romaguera-Ros M, Garcia-Verdugo JM, Aguilar A, Schneider-Maunoury S, Alvarez-Buylla A (2008) Hedgehog signaling and primary cilia are required for the formation of adult neural stem cells. Nat Neurosci 11(3):277–284

Haughey NJ, Liu D, Nath A, Borchard AC, Mattson MP (2002a) Disruption of neurogenesis in the subventricular zone of adult mice, and in human cortical neuronal precursor cells in culture, by amyloid beta-peptide: implications for the pathogenesis of Alzheimer's disease. Neuromolecular Med 1(2):125–135

Haughey NJ, Nath A, Chan SL, Borchard AC, Rao MS, Mattson MP (2002b) Disruption of neurogenesis by amyloid beta-peptide, and perturbed neural progenitor cell homeostasis, in models of Alzheimer's disease. J Neurochem 83(6):1509–1524

Hayes NL, Nowakowski RS (2002) Dynamics of cell proliferation in the adult dentate gyrus of two inbred strains of mice. Brain Res Dev Brain Res 134(1–2):77–85

Hesselager G, Uhrbom L, Westermark B, Nister M (2003) Complementary effects of platelet-derived growth factor autocrine stimulation and p53 or Ink4a–Arf deletion in a mouse glioma model. Cancer Res 63(15):4305–4309

Hopewell JW (1975) The subependymal plate and the genesis of gliomas. J Pathol 117(2):101–103

Horner PJ, Gage FH (2000) Regenerating the damaged central nervous system. Nature 407(6807):963–970

Horner PJ, Power AE, Kempermann G, Kuhn HG, Palmer TD, Winkler J, Thal LJ, Gage FH (2000) Proliferation and differentiation of progenitor cells throughout the intact adult rat spinal cord. J Neurosci 20(6):2218–2228

Iwai M, Sato K, Omori N, Nagano I, Manabe Y, Shoji M, Abe K (2002) Three steps of neural stem cells development in gerbil dentate gyrus after transient ischemia. J Cereb Blood Flow Metab 22(4):411–419

Jackson EL, Garcia-Verdugo JM, Gil-Perotin S, Roy M, Quinones-Hinojosa A, VandenBerg S, Alvarez-Buylla A (2006) PDGFR alpha-positive B cells are neural stem cells in the adult SVZ that form glioma-like growths in response to increased PDGF signaling. Neuron 51(2):187–199

Jankovski A, Sotelo C (1996) Subventricular zone-olfactory bulb migratory pathway in the adult mouse: cellular composition and specificity as determined by heterochronic and heterotopic transplantation. J Comp Neurol 371(3):376–396

Jiang W, Gu W, Brannstrom T, Rosqvist R, Wester P (2001) Cortical neurogenesis in adult rats after transient middle cerebral artery occlusion. Stroke 32(5):1201–1207

Jin K, Minami M, Lan JQ, Mao XO, Batteur S, Simon RP, Greenberg DA (2001) Neurogenesis in dentate subgranular zone and rostral subventricular zone after focal cerebral ischemia in the rat. Proc Natl Acad Sci USA 98(8):4710–4715

Jin K, Mao XO, Sun Y, Xie L, Jin L, Nishi E, Klagsbrun M, Greenberg DA (2002a) Heparin-binding epidermal growth factor-like growth factor: hypoxia-inducible expression in vitro and stimulation of neurogenesis in vitro and in vivo. J Neurosci 22(13):5365–5373

Jin K, Zhu Y, Sun Y, Mao XO, Xie L, Greenberg DA (2002b) Vascular endothelial growth factor (VEGF) stimulates neurogenesis in vitro and in vivo. Proc Natl Acad Sci USA 99(18):11946–11950

Jin K, Sun Y, Xie L, Childs J, Mao XO, Greenberg DA (2004) Post-ischemic administration of heparin-binding epidermal growth factor-like growth factor (HB-EGF) reduces infarct size and modifies neurogenesis after focal cerebral ischemia in the rat. J Cereb Blood Flow Metab 24(4):399–408

Johansson CB, Momma S, Clarke DL, Risling M, Lendahl U, Frisen J (1999) Identification of a neural stem cell in the adult mammalian central nervous system. Cell 96(1):25–34

Jordan JD, Ming GL, Song H (2006) Adult neurogenesis as a potential therapy for neurodegenerative diseases. Discov Med 6(34):144–147

Jorgensen HS, Plesner AM, Hubbe P, Larsen K (1992) Marked increase of stroke incidence in men between 1972 and 1990 in Frederiksberg, Denmark. Stroke 23(12):1701–1704

Kalyani AJ, Piper D, Mujtaba T, Lucero MT, Rao MS (1998) Spinal cord neuronal precursors generate multiple neuronal phenotypes in culture. J Neurosci 18(19):7856–7868

Kaneko Y, Sakakibara S, Imai T, Suzuki A, Nakamura Y, Sawamoto K, Ogawa Y, Toyama Y, Miyata T, Okano H (2000) Musashi1: an evolutionarily conserved marker for CNS progenitor cells including neural stem cells. Dev Neurosci 22(1–2):139–153

Kaplan MS (1981) Neurogenesis in the 3-month-old rat visual cortex. J Comp Neurol 195(2):323–338

Kaplan MS (1985) Formation and turnover of neurons in young and senescent animals: an electronmicroscopic and morphometric analysis. Ann NY Acad Sci 457:173–192

Kaplan MS, Bell DH (1984) Mitotic neuroblasts in the 9-day-old and 11-month-old rodent hippocampus. J Neurosci 4(6):1429–1441

Kaplan MS, McNelly NA, Hinds JW (1985) Population dynamics of adult-formed granule neurons of the rat olfactory bulb. J Comp Neurol 239(1):117–125

Kawaguchi A, Miyata T, Sawamoto K, Takashita N, Murayama A, Akamatsu W, Ogawa M, Okabe M, Tano Y, Goldman SA, Okano H (2001) Nestin-EGFP transgenic mice: visualization of the self-renewal and multipotency of CNS stem cells. Mol Cell Neurosci 17(2):259–273

Kawahara N, Mishima K, Higashiyama S, Taniguchi N, Tamura A, Kirino T (1999) The gene for heparin-binding epidermal growth factor-like growth factor is stress-inducible: its role in cerebral ischemia. J Cereb Blood Flow Metab 19(3):307–320

Kee NJ, Preston E, Wojtowicz JM (2001) Enhanced neurogenesis after transient global ischemia in the dentate gyrus of the rat. Exp Brain Res 136(3):313–320

Kempermann G, Gage FH (2002) Genetic influence on phenotypic differentiation in adult hippocampal neurogenesis. Brain Res Dev Brain Res 134(1–2):1–12

Kempermann G, Kuhn HG, Gage FH (1997) Genetic influence on neurogenesis in the dentate gyrus of adult mice. Proc Natl Acad Sci USA 94(19):10409–10414

Kim M, Morshead CM (2003) Distinct populations of forebrain neural stem and progenitor cells can be isolated using side-population analysis. J Neurosci 23(33):10703–10709

Kirsche W (1967) On postembryonic matrix zones in the brain of various vertebrates and their relationship to the study of the brain structure. Z Mikrosk Anat Forsch 77(3):313–406

Kirschenbaum B, Goldman SA (1995) Brain-derived neurotrophic factor promotes the survival of neurons arising from the adult rat forebrain subependymal zone. Proc Natl Acad Sci USA 92(1):210–214

Koketsu D, Mikami A, Miyamoto Y, Hisatsune T (2003) Nonrenewal of neurons in the cerebral neocortex of adult macaque monkeys. J Neurosci 23(3):937–942

Kornack DR, Rakic P (2001) Cell proliferation without neurogenesis in adult primate neocortex. Science 294(5549):2127–2130

Kosaka T, Hama K (1986) Three-dimensional structure of astrocytes in the rat dentate gyrus. J Comp Neurol 249(2):242–260

Kriss JP, Revesz L (1961) Quantitative studies of incorporation of exogenous thymidine and 5-bromodeoxyuridine into deoxyribonucleic acid of mammalian cells in vitro. Cancer Res 21:1141–1147

Kuhn HG, Winkler J, Kempermann G, Thal LJ, Gage FH (1997) Epidermal growth factor and fibroblast growth factor-2 have different effects on neural progenitors in the adult rat brain. J Neurosci 17(15):5820–5829

Kuo CT, Mirzadeh Z, Soriano-Navarro M, Rasin M, Wang D, Shen J, Sestan N, Garcia-Verdugo J, Alvarez-Buylla A, Jan LY, Jan YN (2006) Postnatal deletion of numb/numblike reveals repair and remodeling capacity in the subventricular neurogenic niche. Cell 127(6):1253–1264

Kurki P, Vanderlaan M, Dolbeare F, Gray J, Tan EM (1986) Expression of proliferating cell nuclear antigen (PCNA)/cyclin during the cell cycle. Exp Cell Res 166(1):209–219

Lagasse E, Connors H, Al-Dhalimy M, Reitsma M, Dohse M, Osborne L, Wang X, Finegold M, Weissman IL, Grompe M (2000) Purified hematopoietic stem cells can differentiate into hepatocytes in vivo. Nat Med 6(11):1229–1234

Lago N, Ceballos D, Rodriguez FJ, Stieglitz T, Navarro X (2005) Long term assessment of axonal regeneration through polyimide regenerative electrodes to interface the peripheral nerve. Biomaterials 26(14):2021–2031

Lantos PL, Roscoe JP, Skidmore CJ (1976) Studies of the morphology and tumorigenicity of experimental brain tumours in tissue culture. Br J Exp Pathol 57(1):95–104

Lendahl U, Zimmerman LB, McKay RD (1990) CNS stem cells express a new class of intermediate filament protein. Cell 60(4):585–595

Leonhardt H (1972) [Electron microscopy studies on the past embryonic ventral matrix zone in the rabbit brain]. Zeitschrift fur mikroskopisch-anatomische Forschung 85(2):161–175

Leonard JR, D'Sa C, Klocke BJ, Roth KA (2001) Neural precursor cell apoptosis and glial tumorigenesis following transplacental ethyl-nitrosourea exposure. Oncogene 20(57):8281–8286

Levitt P, Cooper ML, Rakic P (1981) Coexistence of neuronal and glial precursor cells in the cerebral ventricular zone of the fetal monkey: an ultrastructural immunoperoxidase analysis. J Neurosci 1(1):27–39

Lewis PD (1968) Mitotic activity in the primate subependymal layer and the genesis of gliomas. Nature 217(132):974–975

Li J, Yen C, Liaw D, Podsypanina K, Bose S, Wang SI, Puc J, Miliaresis C, Rodgers L, McCombie R, Bigner SH, Giovanella BC, Ittmann M, Tycko B, Hibshoosh H, Wigler MH, Parsons R (1997) PTEN, a putative protein tyrosine phosphatase gene mutated in human brain, breast, and prostate cancer. Science 275(5308):1943–1947

Li L, Liu F, Salmonsen RA, Turner TK, Litofsky NS, Di Cristofano A, Pandolfi PP, Jones SN, Recht LD, Ross AH (2002) PTEN in neural precursor cells: regulation of migration, apoptosis, and proliferation. Mol Cell Neurosci 20(1):21–29

Lie DC, Song H, Colamarino SA, Ming GL, Gage FH (2004) Neurogenesis in the adult brain: new strategies for central nervous system diseases. Annu Rev Pharmacol Toxicol 44:399–421

Lillien L, Raphael H (2000) BMP and FGF regulate the development of EGF-responsive neural progenitor cells. Development 127(22):4993–5005

Lim DA, Tramontin AD, Trevejo JM, Herrera DG, Garcia-Verdugo JM, Alvarez-Buylla A (2000) Noggin antagonizes BMP signaling to create a niche for adult neurogenesis. Neuron 28(3):713–726

Lindvall O, Kokaia Z (2006) Stem cells for the treatment of neurological disorders. Nature 441(7097):1094–1096

Liu J, Solway K, Messing RO, Sharp FR (1998) Increased neurogenesis in the dentate gyrus after transient global ischemia in gerbils. J Neurosci 18(19):7768–7778

Lois C, Alvarez-Buylla A (1993) Proliferating subventricular zone cells in the adult mammalian forebrain can differentiate into neurons and glia. Proc Natl Acad Sci USA 90(5):2074–2077

Lois C, Garcia-Verdugo JM, Alvarez-Buylla A (1996) Chain migration of neuronal precursors. Science 271(5251):978–981

Lopez-Garcia C, Molowny A, Garcia-Verdugo JM, Ferrer I (1988) Delayed postnatal neurogenesis in the cerebral cortex of lizards. Brain Res 471(2):167–174

Lopez-Garcia C, Molowny A, Martinez-Guijarro FJ, Blasco-Ibanez JM, Luis de la Iglesia JA, Bernabeu A, Garcia-Verdugo JM (1992) Lesion and regeneration in the medial cerebral cortex of lizards. Histol Histopathol 7(4):725–746

Luskin MB (1993) Restricted proliferation and migration of postnatally generated neurons derived from the forebrain subventricular zone. Neuron 11(1):173–189

Luzzati F, Peretto P, Aimar P, Ponti G, Fasolo A, Bonfanti L (2003) Glia-independent chains of neuroblasts through the subcortical parenchyma of the adult rabbit brain. Proc Natl Acad Sci USA 100(22):13036–13041

Luzzati F, De Marchis S, Fasolo A, Peretto P (2006) Neurogenesis in the caudate nucleus of the adult rabbit. J Neurosci 26(2):609–621

Magavi SS, Leavitt BR, Macklis JD (2000) Induction of neurogenesis in the neocortex of adult mice. Nature 405(6789):951–955

Magrassi L, Castello S, Ciardelli L, Podesta M, Gasparoni A, Conti L, Pezzotta S, Frassoni F, Cattaneo E (2003) Freshly dissociated fetal neural stem/progenitor cells do not turn into blood. Mol Cell Neurosci 22(2):179–187

Maric D, Barker JL (2004) Neural stem cells redefined: a FACS perspective. Mol Neurobiol 30(1):49–76

Markakis EA, Gage FH (1999) Adult-generated neurons in the dentate gyrus send axonal projections to field CA3 and are surrounded by synaptic vesicles. J Comp Neurol 406(4):449–460

Marti HH (2004) Erythropoietin and the hypoxic brain. J Exp Biol 207(Pt 18):3233–3242

Martino G, Pluchino S (2006) The therapeutic potential of neural stem cells. Nat Rev 7(5):395–406

Matsuoka N, Nozaki K, Takagi Y, Nishimura M, Hayashi J, Miyatake S, Hashimoto N (2003) Adenovirus-mediated gene transfer of fibroblast growth factor-2 increases BrdU-positive cells after forebrain ischemia in gerbils. Stroke 34(6):1519–1525

McDermott KW, Lantos PL (1990) Cell proliferation in the subependymal layer of the postnatal marmoset, Callithrix jacchus. Brain Res Dev Brain Res 57(2):269–277

McDonald JW (2004) Repairing the damaged spinal cord: from stem cells to activity-based restoration therapies. Clin Neurosurg 51:207–227

Menezes JR, Luskin MB (1994) Expression of neuron-specific tubulin defines a novel population in the proliferative layers of the developing telencephalon. J Neurosci 14(9):5399–5416

Menn B, Garcia-Verdugo JM, Yaschine C, Gonzalez-Perez O, Rowitch D, Alvarez-Buylla A (2006) Origin of oligodendrocytes in the subventricular zone of the adult brain. J Neurosci 26(30):7907–7918

Merkle FT, Tramontin AD, Garcia-Verdugo JM, Alvarez-Buylla A (2004) Radial glia give rise to adult neural stem cells in the subventricular zone. Proc Natl Acad Sci USA 101(50):17528–17532

Micheva KD, Smith SJ (2007) Array tomography: a new tool for imaging the molecular architecture and ultrastructure of neural circuits. Neuron 55(1):25–36

Mitro A, Palkovits M (1981) Morphology of the rat brain ventricles, ependyma, and periventricular structures. Bibl Anat 21:1–110

Mizumatsu S, Monje ML, Morhardt DR, Rola R, Palmer TD, Fike JR (2003) Extreme sensitivity of adult neurogenesis to low doses of X-irradiation. Cancer Res 63(14):4021–4027

Morshead CM, Benveniste P, Iscove NN, van der Kooy D (2002) Hematopoietic competence is a rare property of neural stem cells that may depend on genetic and epigenetic alterations. Nat Med 8(3):268–273

Murayama A, Matsuzaki Y, Kawaguchi A, Shimazaki T, Okano H (2002) Flow cytometric analysis of neural stem cells in the developing and adult mouse brain. J Neurosci Res 69(6):837–847

Nacher J, Crespo C, McEwen BS (2001) Doublecortin expression in the adult rat telencephalon. Eur J Neurosci 14(4):629–644

Nakatomi H, Kuriu T, Okabe S, Yamamoto S, Hatano O, Kawahara N, Tamura A, Kirino T, Nakafuku M (2002) Regeneration of hippocampal pyramidal neurons after ischemic brain injury by recruitment of endogenous neural progenitors. Cell 110(4):429–441

Navarro-Quiroga I, Hernandez-Valdes M, Lin SL, Naegele JR (2006) Postnatal cellular contributions of the hippocampus subventricular zone to the dentate gyrus, corpus callosum, fimbria, and cerebral cortex. J Comp Neurol 497(5):833–845

Nieto-Sampedro M (2003) Central nervous system lesions that can and those that cannot be repaired with the help of olfactory bulb ensheathing cell transplants. Neurochem Res 28(11):1659–1676

Nottebohm F (1985) Neuronal replacement in adulthood. Ann NY Acad Sci 457:143–161

Novikov LN, Novikova LN, Mosahebi A, Wiberg M, Terenghi G, Kellerth JO (2002) A novel biodegradable implant for neuronal rescue and regeneration after spinal cord injury. Biomaterials 23(16):3369–3376

Nowakowski RS, Lewin SB, Miller MW (1989) Bromodeoxyuridine immunohistochemical determination of the lengths of the cell cycle and the DNA-synthetic phase for an anatomically defined population. J Neurocytol 18(3):311–318

Nunes MC, Roy NS, Keyoung HM, Goodman RR, McKhann G II, Jiang L, Kang J, Nedergaard M, Goldman SA (2003) Identification and isolation of multipotential neural progenitor cells from the subcortical white matter of the adult human brain. Nat Med 9(4):439–447

Oda H, Zhang S, Tsurutani N, Shimizu S, Nakatsuru Y, Aizawa S, Ishikawa T (1997) Loss of p53 is an early event in induction of brain tumors in mice by transplacental carcinogen exposure. Cancer Res 57(4):646–650

Ong J, Plane JM, Parent JM, Silverstein FS (2005) Hypoxic–ischemic injury stimulates subventricular zone proliferation and neurogenesis in the neonatal rat. Pediatr Res 58(3):600–606

Palma V, Lim DA, Dahmane N, Sanchez P, Brionne TC, Herzberg CD, Gitton Y, Carleton A, Alvarez-Buylla A, Ruiz i Altaba A (2005) Sonic hedgehog controls stem cell behavior in the postnatal and adult brain. Development 132(2):335–344

Palmer TD, Willhoite AR, Gage FH (2000) Vascular niche for adult hippocampal neurogenesis. J Comp Neurol 425(4):479–494

Pardal R, Molofsky AV, He S, Morrison SJ (2005) Stem cell self-renewal and cancer cell proliferation are regulated by common networks that balance the activation of proto-oncogenes and tumor suppressors. Cold Spring Harbor Symp Quant Biol 70:177–185

Parent JM, Yu TW, Leibowitz RT, Geschwind DH, Sloviter RS, Lowenstein DH (1997) Dentate granule cell neurogenesis is increased by seizures and contributes to aberrant network reorganization in the adult rat hippocampus. J Neurosci 17(10):3727–3738

Parent JM, Valentin VV, Lowenstein DH (2002) Prolonged seizures increase proliferating neuroblasts in the adult rat subventricular zone-olfactory bulb pathway. J Neurosci 22(8):3174–3188

Parent JM, von dem Bussche N, Lowenstein DH (2006) Prolonged seizures recruit caudal subventricular zone glial progenitors into the injured hippocampus. Hippocampus 16(3):321–328

Parker MA, Anderson JK, Corliss DA, Abraria VE, Sidman RL, Park KI, Teng YD, Cotanche DA, Snyder EY (2005) Expression profile of an operationally-defined neural stem cell clone. Exp Neurol 194(2):320–332

Paton JA, Nottebohm FN (1984) Neurons generated in the adult brain are recruited into functional circuits. Science 225(4666):1046–1048

Pencea V, Bingaman KD, Freedman LJ, Luskin MB (2001a) Neurogenesis in the subventricular zone and rostral migratory stream of the neonatal and adult primate forebrain. Exp Neurol 172(1):1–16

Pencea V, Bingaman KD, Wiegand SJ, Luskin MB (2001b) Infusion of brain-derived neurotrophic factor into the lateral ventricle of the adult rat leads to new neurons in the parenchyma of the striatum, septum, thalamus, and hypothalamus. J Neurosci 21(17):6706–6717

Perego C, Vanoni C, Massari S, Raimondi A, Pola S, Cattaneo MG, Francolini M, Vicentini LM, Pietrini G (2002) Invasive behaviour of glioblastoma cell lines is associated with altered organisation of the cadherin–catenin adhesion system. J Cell Sci 115(Pt 16):3331–3340

Peretto P, Bonfanti L, Merighi A, Fasolo A (1998) Carnosine-like immunoreactivity in astrocytes of the glial tubes and in newly-generated cells within the tangential part of the rostral migratory stream of rodents. Neuroscience 85(2):527–542

Perez-Martin M, Grondona JM, Cifuentes M, Perez-Figares JM, Jimenez JA, Fernandez-Llebrez P (2000) Ependymal explants from the lateral ventricle of the adult bovine brain: a model system for morphological and functional studies of the ependyma. Cell Tissue Res 300(1):11–19

Perez-Martin M, Cifuentes M, Grondona JM, Bermudez-Silva FJ, Arrabal PM, Perez-Figares JM, Jimenez AJ, Garcia-Segura LM, Fernandez-Llebrez P (2003) Neurogenesis in explants from the walls of the lateral ventricle of adult bovine brain: role of endogenous IGF-1 as a survival factor. Eur J Neurosci 17(2):205–211

Pfenninger CV, Roschupkina T, Hertwig F, Kottwitz D, Englund E, Bengzon J, Jacobsen SE, Nuber UA (2007) CD133 is not present on neurogenic astrocytes in the adult subventricular zone, but on embryonic neural stem cells, ependymal cells, and glioblastoma cells. Cancer Res 67(12):5727–5736

Plate KH, Breier G, Weich HA, Mennel HD, Risau W (1994) Vascular endothelial growth factor and glioma angiogenesis: coordinate induction of VEGF receptors, distribution of VEGF protein and possible in vivo regulatory mechanisms. International journal of cancer 59(4):520–529

Pleasure SJ, Collins AE, Lowenstein DH (2000) Unique expression patterns of cell fate molecules delineate sequential stages of dentate gyrus development. J Neurosci 20(16):6095–6105

Pluchino S, Quattrini A, Brambilla E, Gritti A, Salani G, Dina G, Galli R, Del Carro U, Amadio S, Bergami A, Furlan R, Comi G, Vescovi AL, Martino G (2003) Injection of adult neurospheres induces recovery in a chronic model of multiple sclerosis. Nature 422(6933):688–694

Pluchino S, Zanotti L, Martino G (2007) Rationale for the use of neural stem/precursor cells in immune-mediated demyelinating disorders. J Neurol 254(Suppl 1):I23–I28

Politi LS, Bacigaluppi M, Brambilla E, Cadioli M, Falini A, Comi G, Scotti G, Martino G, Pluchino S (2007) Magnetic-resonance-based tracking and quantification of intravenously injected neural stem cell accumulation in the brains of mice with experimental multiple sclerosis. Stem Cells 25(10):2583–2592

Ponti G, Aimar P, Bonfanti L (2006) Cellular composition and cytoarchitecture of the rabbit subventricular zone and its extensions in the forebrain. J Comp Neurol 498(4):491–507

Potten CS, Loeffler M (1990) Stem cells: attributes, cycles, spirals, pitfalls and uncertainties. Lessons for and from the crypt. Development 110(4):1001–1020

Quinones-Hinojosa A, Sanai N, Soriano-Navarro M, Gonzalez-Perez O, Mirzadeh Z, Gil-Perotin S, Romero-Rodriguez R, Berger MS, Garcia-Verdugo JM, Alvarez-Buylla A (2006) Cellular composition and cytoarchitecture of the adult human subventricular zone: a niche of neural stem cells. J Comp Neurol 494(3):415–434

Rakic P (1985a) DNA synthesis and cell division in the adult primate brain. Ann NY Acad Sci 457:193–211

Rakic P (1985b) Limits of neurogenesis in primates. Science 227(4690):1054–1056

Ramon y Cajal S (1913) Estudios sobre la degeneración y regeneración. Imprenta de Hijos de Nicolás Moya, Madrid

Ramon-Cueto A, Cordero MI, Santos-Benito FF, Avila J (2000) Functional recovery of paraplegic rats and motor axon regeneration in their spinal cords by olfactory ensheathing glia. Neuron 25(2):425–435

Reilly KM, Tuskan RG, Christy E, Loisel DA, Ledger J, Bronson RT, Smith CD, Tsang S, Munroe DJ, Jacks T (2004) Susceptibility to astrocytoma in mice mutant for Nf1 and Trp53 is linked to chromosome 11 and subject to epigenetic effects. Proc Natl Acad Sci USA 101(35):13008–13013

Reynolds BA, Weiss S (1992) Generation of neurons and astrocytes from isolated cells of the adult mammalian central nervous system. Science 255(5052):1707–1710

Reynolds BA, Tetzlaff W, Weiss S (1992) A multipotent EGF-responsive striatal embryonic progenitor cell produces neurons and astrocytes. J Neurosci 12(11):4565–4574

Rietze R, Poulin P, Weiss S (2000) Mitotically active cells that generate neurons and astrocytes are present in multiple regions of the adult mouse hippocampus. J Comp Neurol 424(3):397–408

Rietze RL, Valcanis H, Brooker GF, Thomas T, Voss AK, Bartlett PF (2001) Purification of a pluripotent neural stem cell from the adult mouse brain. Nature 412(6848):736–739

Risau W (1997) Mechanisms of angiogenesis. Nature 386(6626):671–674

Rodic N, Rutenberg MS, Terada N (2004) Cell fusion and reprogramming: resolving our transdifferences. Trends Mol Med 10(3):93–96

Rodriguez-Perez LM, Perez-Martin M, Jimenez AJ, Fernandez-Llebrez P (2003) Immunocytochemical characterisation of the wall of the bovine lateral ventricle. Cell Tissue Res 314(3):325–335

Rola R, Mizumatsu S, Otsuka S, Morhardt DR, Noble-Haeusslein LJ, Fishman K, Potts MB, Fike JR (2006) Alterations in hippocampal neurogenesis following traumatic brain injury in mice. Exp Neurol 202(1):189–199

Rousselot P, Lois C, Alvarez-Buylla A (1995) Embryonic (PSA) N-CAM reveals chains of migrating neuroblasts between the lateral ventricle and the olfactory bulb of adult mice. J Comp Neurol 351(1):51–61

Roy NS, Benraiss A, Wang S, Fraser RA, Goodman R, Couldwell WT, Nedergaard M, Kawaguchi A, Okano H, Goldman SA (2000a) Promoter-targeted selection and isolation of neural progenitor cells from the adult human ventricular zone. J Neurosci Res 59(3):321–331

Roy NS, Wang S, Jiang L, Kang J, Benraiss A, Harrison-Restelli C, Fraser RA, Couldwell WT, Kawaguchi A, Okano H, Nedergaard M, Goldman SA (2000b) In vitro neurogenesis by progenitor cells isolated from the adult human hippocampus. Nat Med 6(3):271–277

Russo RE, Fernandez A, Reali C, Radmilovich M, Trujillo-Cenoz O (2004) Functional and molecular clues reveal precursor-like cells and immature neurones in the turtle spinal cord. J Physiol 560(Pt 3):831–838

Sakakibara S, Okano H (1997) Expression of neural RNA-binding proteins in the postnatal CNS: implications of their roles in neuronal and glial cell development. J Neurosci 17(21):8300–8312

Sakakibara S, Imai T, Hamaguchi K, Okabe M, Aruga J, Nakajima K, Yasutomi D, Nagata T, Kurihara Y, Uesugi S, Miyata T, Ogawa M, Mikoshiba K, Okano H (1996) Mouse-Musashi-1, a neural RNA-binding protein highly enriched in the mammalian CNS stem cell. Dev Biol 176(2):230–242

Sanai N, Tramontin AD, Quinones-Hinojosa A, Barbaro NM, Gupta N, Kunwar S, Lawton MT, McDermott MW, Parsa AT, Manuel-Garcia Verdugo J, Berger MS, Alvarez-Buylla A (2004) Unique astrocyte ribbon in adult human brain contains neural stem cells but lacks chain migration. Nature 427(6976):740–744

Sanai N, Berger MS, Garcia-Verdugo JM, Alvarez-Buylla A (2007) Comment on "Human neuroblasts migrate to the olfactory bulb via a lateral ventricular extension". Science 318(5849):393; author reply 393

Saravia F, Revsin Y, Lux-Lantos V, Beauquis J, Homo-Delarche F, De Nicola AF (2004) Oestradiol restores cell proliferation in dentate gyrus and subventricular zone of streptozotocin-diabetic mice. J Neuroendocrinol 16(8):704–710

Sauer B (1998) Inducible gene targeting in mice using the Cre/lox system. Methods 14(4):381–392

Sawamoto K, Wichterle H, Gonzalez-Perez O, Cholfin JA, Yamada M, Spassky N, Murcia NS, Garcia-Verdugo JM, Marin O, Rubenstein JL, Tessier-Lavigne M, Okano H, Alvarez-Buylla A (2006) New neurons follow the flow of cerebrospinal fluid in the adult brain. Science 311(5761):629–632

Schmidt W, Reymann KG (2002) Proliferating cells differentiate into neurons in the hippocampal CA1 region of gerbils after global cerebral ischemia. Neurosci Lett 334(3):153–156

Schneider T, Sailer M, Ansorge S, Firsching R, Reinhold D (2006) Increased concentrations of transforming growth factor beta1 and beta2 in the plasma of patients with glioblastoma. Journal of neuro-oncology 79(1):61–65

Seri B, Garcia-Verdugo JM, McEwen BS, Alvarez-Buylla A (2001) Astrocytes give rise to new neurons in the adult mammalian hippocampus. J Neurosci 21(18):7153–7160

Seri B, Garcia-Verdugo JM, Collado-Morente L, McEwen BS, Alvarez-Buylla A (2004) Cell types, lineage, and architecture of the germinal zone in the adult dentate gyrus. J Comp Neurol 478(4):359–378

Seri B, Herrera DG, Gritti A, Ferron S, Collado L, Vescovi A, Garcia-Verdugo JM, Alvarez-Buylla A (2006) Composition and organization of the SCZ: a large germinal layer containing neural stem cells in the adult mammalian brain. Cereb Cortex 16(Suppl 1):i103–i111

Shapiro EM, Gonzalez-Perez O, Manuel Garcia-Verdugo J, Alvarez-Buylla A, Koretsky AP (2006) Magnetic resonance imaging of the migration of neuronal precursors generated in the adult rodent brain. Neuroimage 32(3):1150–1157

Shibuya S, Miyamoto O, Itano T, Mori S, Norimatsu H (2003) Temporal progressive antigen expression in radial glia after contusive spinal cord injury in adult rats. Glia 42(2):172–183

Shihabuddin LS, Ray J, Gage FH (1997) FGF-2 is sufficient to isolate progenitors found in the adult mammalian spinal cord. Exp Neurol 148(2):577–586

Shihabuddin LS, Horner PJ, Ray J, Gage FH (2000) Adult spinal cord stem cells generate neurons after transplantation in the adult dentate gyrus. J Neurosci 20(23):8727–8735

Shimazaki T, Shingo T, Weiss S (2001) The ciliary neurotrophic factor/leukemia inhibitory factor/gp130 receptor complex operates in the maintenance of mammalian forebrain neural stem cells. J Neurosci 21(19):7642–7653

Shingo T, Sorokan ST, Shimazaki T, Weiss S (2001) Erythropoietin regulates the in vitro and in vivo production of neuronal progenitors by mammalian forebrain neural stem cells. J Neurosci 21(24):9733–9743

Shors TJ, Miesegaes G, Beylin A, Zhao M, Rydel T, Gould E (2001) Neurogenesis in the adult is involved in the formation of trace memories. Nature 410(6826):372–376

Shoshan Y, Nishiyama A, Chang A, Mork S, Barnett GH, Cowell JK, Trapp BD, Staugaitis SM (1999) Expression of oligodendrocyte progenitor cell antigens by gliomas: implications for the histogenesis of brain tumors. Proc Natl Acad Sci USA 96(18):10361–10366

Shyu WC, Lin SZ, Yang HI, Tzeng YS, Pang CY, Yen PS, Li H (2004) Functional recovery of stroke rats induced by granulocyte colony-stimulating factor-stimulated stem cells. Circulation 110(13):1847–1854

Singh SK, Clarke ID, Terasaki M, Bonn VE, Hawkins C, Squire J, Dirks PB (2003) Identification of a cancer stem cell in human brain tumors. Cancer Res 63(18):5821–5828

Singla V, Reiter JF (2006) The primary cilium as the cell's antenna: signaling at a sensory organelle. Science 313(5787):629–633

Song H, Stevens CF, Gage FH (2002) Astroglia induce neurogenesis from adult neural stem cells. Nature 417(6884):39–44

Sonntag KC, Simantov R, Isacson O (2005) Stem cells may reshape the prospect of Parkinson's disease therapy. Brain Res Mol Brain Res 134(1):34–51

Spassky N, Han YG, Aguilar A, Strehl L, Besse L, Laclef C, Ros MR, Garcia-Verdugo JM, Alvarez-Buylla A (2008) Primary cilia are required for cerebellar development and Shh-dependent expansion of progenitor pool. Dev Biol 317(1):246–259

References

Stanfield BB, Trice JE (1988) Evidence that granule cells generated in the dentate gyrus of adult rats extend axonal projections. Exp Brain Res 72(2):399–406

Sugita N (1918) Comparative studies on the growth of the cerebral cortex. J Comp Neurol 29:177–241

Suslov ON, Kukekov VG, Ignatova TN, Steindler DA (2002) Neural stem cell heterogeneity demonstrated by molecular phenotyping of clonal neurospheres. Proc Natl Acad Sci USA 99(22):14506–14511

Tabatabai G, Bahr O, Mohle R, Eyupoglu IY, Boehmler AM, Wischhusen J, Rieger J, Blumcke I, Weller M, Wick W (2005) Lessons from the bone marrow: how malignant glioma cells attract adult haematopoietic progenitor cells. Brain 128(Pt 9):2200–2211

Taguchi A, Soma T, Tanaka H, Kanda T, Nishimura H, Yoshikawa H, Tsukamoto Y, Iso H, Fujimori Y, Stern DM, Naritomi H, Matsuyama T (2004) Administration of CD34+ cells after stroke enhances neurogenesis via angiogenesis in a mouse model. J Clin Invest 114(3):330–338

Takasaki Y, Deng JS, Tan EM (1981) A nuclear antigen associated with cell proliferation and blast transformation. J Exp Med 154(6):1899–1909

Tanaka R, Yamashiro K, Mochizuki H, Cho N, Onodera M, Mizuno Y, Urabe T (2004) Neurogenesis after transient global ischemia in the adult hippocampus visualized by improved retroviral vector. Stroke 35(6):1454–1459

Taupin P (2006) The therapeutic potential of adult neural stem cells. Curr Opin Mol Ther 8(3):225–231

Temple S (2001) The development of neural stem cells. Nature 414(6859):112–117

Terada N, Hamazaki T, Oka M, Hoki M, Mastalerz DM, Nakano Y, Meyer EM, Morel L, Petersen BE, Scott EW (2002) Bone marrow cells adopt the phenotype of other cells by spontaneous cell fusion. Nature 416(6880):542–545

Teramoto T, Qiu J, Plumier JC, Moskowitz MA (2003) EGF amplifies the replacement of parvalbumin-expressing striatal interneurons after ischemia. J Clin Invest 111(8):1125–1132

Terent A (1988) Increasing incidence of stroke among Swedish women. Stroke 19(5):598–603

Thomas LB, Gates MA, Steindler DA (1996) Young neurons from the adult subependymal zone proliferate and migrate along an astrocyte, extracellular matrix-rich pathway. Glia 17(1):1–14

Thored P, Arvidsson A, Cacci E, Ahlenius H, Kallur T, Darsalia V, Ekdahl CT, Kokaia Z, Lindvall O (2006) Persistent production of neurons from adult brain stem cells during recovery after stroke. Stem Cells 24(3):739–747

Thored P, Wood J, Arvidsson A, Cammenga J, Kokaia Z, Lindvall O (2007) Long-term neuroblast migration along blood vessels in an area with transient angiogenesis and increased vascularization after stroke. Stroke 38(11):3032–3039

Tonchev AB, Yamashima T, Sawamoto K, Okano H (2005) Enhanced proliferation of progenitor cells in the subventricular zone and limited neuronal production in the striatum and neocortex of adult macaque monkeys after global cerebral ischemia. J Neurosci Res 81(6):776–788

Trojan J, Johnson TR, Rudin SD, Ilan J, Tykocinski ML, Ilan J (1993) Treatment and prevention of rat glioblastoma by immunogenic C6 cells expressing antisense insulin-like growth factor I RNA. Science New York, NY 259(5091):94–97

Uberti D, Piccioni L, Cadei M, Grigolato P, Rotter V, Memo M (2001) p53 is dispensable for apoptosis but controls neurogenesis of mouse dentate gyrus cells following gamma-irradiation. Brain Res Mol Brain Res 93(1):81–89

Uchida N, Buck DW, He D, Reitsma MJ, Masek M, Phan TV, Tsukamoto AS, Gage FH, Weissman IL (2000) Direct isolation of human central nervous system stem cells. Proc Natl Acad Sci USA 97(26):14720–14725

Uchida K, Mukai M, Okano H, Kawase T (2004) Possible oncogenicity of subventricular zone neural stem cells: case report. Neurosurgery 55(4):977–978

Uhrbom L, Hesselager G, Nister M, Westermark B (1998) Induction of brain tumors in mice using a recombinant platelet-derived growth factor B-chain retrovirus. Cancer Res 58(23):5275–5279

Uhrbom L, Hesselager G, Ostman A, Nister M, Westermark B (2000) Dependence of autocrine growth factor stimulation in platelet-derived growth factor-B-induced mouse brain tumor cells. Int J Cancer 85(3):398–406

Utsuki S, Sato Y, Oka H, Tsuchiya B, Suzuki S, Fujii K (2002) Relationship between the expression of E-, N-cadherins and beta-catenin and tumor grade in astrocytomas. J Neurooncol 57(3):187–192

Van Duyne GD (2001) A structural view of Cre–loxP site-specific recombination. Annu Rev Biophys Biomol Struct 30:87–104

Van Kampen JM, Robertson HA (2005) A possible role for dopamine D3 receptor stimulation in the induction of neurogenesis in the adult rat substantia nigra. Neuroscience 136(2):381–386

Vassilopoulos G, Wang PR, Russell DW (2003) Transplanted bone marrow regenerates liver by cell fusion. Nature 422(6934):901–904

Vescovi AL, Reynolds BA, Fraser DD, Weiss S (1993) bFGF regulates the proliferative fate of unipotent (neuronal) and bipotent (neuronal/astroglial) EGF-generated CNS progenitor cells. Neuron 11(5):951–966

Vescovi AL, Gritti A, Galli R, Parati EA (1999) Isolation and intracerebral grafting of nontransformed multipotential embryonic human CNS stem cells. J Neurotrauma 16(8):689–693

Vick NA, Lin MJ, Bigner DD (1977) The role of the subependymal plate in glial tumorigenesis. Acta Neuropathol 40(1):63–71

Visted T, Enger PO, Lund-Johansen M, Bjerkvig R (2003) Mechanisms of tumor cell invasion and angiogenesis in the central nervous system. Front Biosci 8:e289–e304

Wagner JP, Black IB, DiCicco-Bloom E (1999) Stimulation of neonatal and adult brain neurogenesis by subcutaneous injection of basic fibroblast growth factor. J Neurosci 19(14):6006–6016

Wang X, Willenbring H, Akkari Y, Torimaru Y, Foster M, Al-Dhalimy M, Lagasse E, Finegold M, Olson S, Grompe M (2003) Cell fusion is the principal source of bone-marrow-derived hepatocytes. Nature 422(6934):897–901

Wang L, Zhang Z, Wang Y, Zhang R, Chopp M (2004) Treatment of stroke with erythropoietin enhances neurogenesis and angiogenesis and improves neurological function in rats. Stroke 35(7):1732–1737

Wechsler-Reya R, Scott MP (2001) The developmental biology of brain tumors. Annual review of neuroscience 24:385–428

Weiss S, Dunne C, Hewson J, Wohl C, Wheatley M, Peterson AC, Reynolds BA (1996) Multipotent CNS stem cells are present in the adult mammalian spinal cord and ventricular neuroaxis. J Neurosci 16(23):7599–7609

Wong DT, Chou MY, Chang LC, Gallagher GT (1990) Use of intracellular H3 messenger RNA as a marker to determine the proliferation pattern of normal and 7,12-dimethylbenz[a]anthracene-transformed hamster oral epithelium. Cancer Res 50(16):5107–5111

Wu JP, Kuo JS, Liu YL, Tzeng SF (2000) Tumor necrosis factor-alpha modulates the proliferation of neural progenitors in the subventricular/ventricular zone of adult rat brain. Neurosci Lett 292(3):203–206

Xu Y, Tamamaki N, Noda T, Kimura K, Itokazu Y, Matsumoto N, Dezawa M, Ide C (2005) Neurogenesis in the ependymal layer of the adult rat 3rd ventricle. Exp Neurol 192(2):251–264

Yagita Y, Kitagawa K, Ohtsuki T, Takasawa K, Miyata T, Okano H, Hori M, Matsumoto M (2001) Neurogenesis by progenitor cells in the ischemic adult rat hippocampus. Stroke 32(8):1890–1896

Yamamoto S, Nagao M, Sugimori M, Kosako H, Nakatomi H, Yamamoto N, Takebayashi H, Nabeshima Y, Kitamura T, Weinmaster G, Nakamura K, Nakafuku M (2001) Transcription factor expression and Notch-dependent regulation of neural progenitors in the adult rat spinal cord. J Neurosci 21(24):9814–9823

Yamashita T, Ninomiya M, Hernandez Acosta P, Garcia-Verdugo JM, Sunabori T, Sakaguchi M, Adachi K, Kojima T, Hirota Y, Kawase T, Araki N, Abe K, Okano H, Sawamoto K (2006) Subventricular zone-derived neuroblasts migrate and differentiate into mature neurons in the post-stroke adult striatum. J Neurosci 26(24):6627–6636

Yang K, Cepko CL (1996) Flk-1, a receptor for vascular endothelial growth factor (VEGF), is expressed by retinal progenitor cells. J Neurosci 16(19):6089–6099

Yang HK, Sundholm-Peters NL, Goings GE, Walker AS, Hyland K, Szele FG (2004) Distribution of doublecortin expressing cells near the lateral ventricles in the adult mouse brain. J Neurosci Res 76(3):282–295

Yehuda R, Fairman KR, Meyer JS (1989) Enhanced brain cell proliferation following early adrenalectomy in rats. J Neurochem 53(1):241–248

Yoshimura S, Takagi Y, Harada J, Teramoto T, Thomas SS, Waeber C, Bakowska JC, Breakefield XO, Moskowitz MA (2001) FGF-2 regulation of neurogenesis in adult hippocampus after brain injury. Proc Natl Acad Sci USA 98(10):5874–5879

Yusta-Boyo MJ, Gonzalez MA, Pavon N, Martin AB, De La Fuente R, Garcia-Castro J, De Pablo F, Moratalla R, Bernad A, Vicario-Abejon C (2004) Absence of hematopoiesis from transplanted olfactory bulb neural stem cells. Eur J Neurosci 19(3):505–512

Zhang R, Wang Y, Zhang L, Zhang Z, Tsang W, Lu M, Zhang L, Chopp M (2002) Sildenafil (Viagra) induces neurogenesis and promotes functional recovery after stroke in rats. Stroke 33(11):2675–2680

Zhang R, Zhang Z, Zhang C, Zhang L, Robin A, Wang Y, Lu M, Chopp M (2004) Stroke transiently increases subventricular zone cell division from asymmetric to symmetric and increases neuronal differentiation in the adult rat. J Neurosci 24(25):5810–5815

Zhang RL, Zhang ZG, Chopp M (2005) Neurogenesis in the adult ischemic brain: generation, migration, survival, and restorative therapy. Neuroscientist 11(5):408–416

Zhang P, Liu Y, Li J, Kang Q, Tian Y, Chen X, Shi Q, Song T (2006) Cell proliferation in ependymal/subventricular zone and nNOS expression following focal cerebral ischemia in adult rats. Neurol Res 28(1):91–96

Zhao M, Momma S, Delfani K, Carlen M, Cassidy RM, Johansson CB, Brismar H, Shupliakov O, Frisen J, Janson AM (2003) Evidence for neurogenesis in the adult mammalian substantia nigra. Proc Natl Acad Sci USA 100(13):7925–7930

Zhu DY, Liu SH, Sun HS, Lu YM (2003) Expression of inducible nitric oxide synthase after focal cerebral ischemia stimulates neurogenesis in the adult rodent dentate gyrus. J Neurosci 23(1):223–229

Zhu Y, Guignard F, Zhao D, Liu L, Burns DK, Mason RP, Messing A, Parada LF (2005) Early inactivation of p53 tumor suppressor gene cooperating with NF1 loss induces malignant astrocytoma. Cancer Cell 8(2):119–130

Zigova T, Pencea V, Wiegand SJ, Luskin MB (1998) Intraventricular administration of BDNF increases the number of newly generated neurons in the adult olfactory bulb. Mol Cell Neurosci 11(4):234–245

Ziu M, Schmidt NO, Cargioli TG, Aboody KS, Black PM, Carroll RS (2006) Glioma-produced extracellular matrix influences brain tumor tropism of human neural stem cells. J Neurooncol 79(2):125–133

Index

A
Adult neurogenesis, 1–25, 27–62, 67–75, 81–83
AraC. *See* Cytosine-beta-D-arabinofuranoside

B
BrdU. *See* Bromodeoxyuridine
Bromodeoxyuridine (BrdU), 4, 6, 7, 9, 25, 49, 50, 56, 58, 59, 72, 82

C
Cancer, 4, 9, 58, 66
Cytosine-beta-D-arabinofuranoside (AraC), 47, 55

E
Electron microscopy, 2–7, 11–17, 22, 27, 29, 30, 34, 38, 44, 45, 50, 61, 65, 71, 81, 82

G
GFAP. *See* Glial fibrillary acidic protein
Glial fibrillary acidic protein (GFAP), 9, 10, 12, 34, 43, 44, 47, 49, 52, 55, 56, 58, 59, 61, 72

H
Hippocampus, 2, 3, 11, 27, 29, 42–45, 48–50, 52, 58, 69
Human, 1, 3–6, 18–20, 38, 47, 49, 53–63, 66, 68, 78, 82, 83

I
Ischemia, 25, 67–75

L
Lateral ventricles, 2, 4, 11, 27, 30, 47, 48, 51, 53–62, 65, 66

M
Methodologies, 5–25

N
Neural stem cell, 1, 7, 9, 11, 12, 19–21, 23, 24, 29, 43, 47–49, 52, 55, 61, 65, 66, 73, 77–79, 81–83
Neurogenic niche, 43, 45–47, 63, 64
Neurosphere, 17–23, 35, 50, 52, 61, 82

O
OB. *See* Olfactory bulb
Olfactory bulb (OB), 2, 3, 10, 12, 23, 25, 27, 29, 33, 40, 42, 48, 51, 56, 61, 66, 72, 81, 82

P
PDGF. *See* Platelet-derived growth factor
Platelet-derived growth factor (PDGF), 48, 64–66
Polysialylated neural cell adhesion molecule (PSA-NCAM), 30, 38, 44, 45, 48, 55, 56, 58
Proliferating cell nuclear antigen, 5, 7, 55, 58
PSA-NCAM. *See* Polysialylated neural cell adhesion molecule

R
RMS. *See* Rostral migratory stream
Rodents, 38, 53, 55, 56, 59, 61, 74, 82

Rostral migratory stream (RMS), 11, 27, 29, 30, 33–37, 40, 42, 44, 50, 55, 56, 58, 61, 72, 81, 82

S
Subventricular zone (SVZ), 6, 8–12, 16–20, 22, 23, 27–51, 53, 55, 56, 58, 59, 61–67, 69–73, 75, 81, 82

SVZ. *See* Subventricular zone

T
Therapy, 50, 75, 77, 79, 82, 83
Tritiated thymidine, 2–4, 6, 7, 16, 25, 44, 47, 49, 50, 82

Printing: Krips bv, Meppel, The Netherlands
Binding: Stürtz, Würzburg, Germany